SOLUTIONS MANUAL TO ACCOMPANY
MODELS FOR LIFE

SOLUTIONS MANUAL TO ACCOMPANY MODELS FOR LIFE: AN INTRODUCTION TO DISCRETE MATHEMATICAL MODELING WITH MICROSOFT® OFFICE EXCEL®

JEFFREY T. BARTON
Birmingham-Southern College
Birmingham, Alabama, USA

WILEY

Published by John Wiley & Sons, Inc., Hoboken, New Jersey

Published simultaneously in Canada

For general information on our other products and services or for technical support, please contact our Customer Care Department within the United States at (800) 762-2974, outside the United States at (317) 572-3993 or fax (317) 572-4002.

Wiley also publishes its books in a variety of electronic formats. Some content that appears in print may not be available in electronic formats. For more information about Wiley products, visit our web site at www.wiley.com.

Library of Congress Cataloging-in-Publication Data

Names: Barton, Jeffrey T., 1970–
Title: Models for life : an introduction to discrete mathematical modeling with Excel / Jeffrey T. Barton.
Description: Hoboken, New Jersey : John Wiley & Sons, Inc., [2016] | "Excel" in title followed by trademark symbol. | Includes bibliographical references.
Identifiers: LCCN 2015036805 | ISBN 9781119039754 (cloth)
Subjects: LCSH: Differentiable dynamical systems. | Dynamic programming. | Microsoft Excel (Computer file)
Classification: LCC QA614.8 .B3774 2016 | DDC 511/.10285554–dc23 LC record available at http://lccn.loc.gov/2015036805

Set in 10/12pt Times by SPi Global, Pondicherry, India

10 9 8 7 6 5 4 3 2 1

1   2016

CONTENTS

PREFACE

Included in this manual are complete, step-by-step solutions to all odd-numbered problems from the text *Models for Life: An Introduction to Discrete Mathematical Modeling with Microsoft® Office Excel®*. As is the case in many areas of mathematics, in mathematical modeling there are often several reasonable ways to approach a problem, and an effort has been made in this manual to mention alternate approaches to problems where appropriate. Please keep in mind that if a proposed solution does not exactly match the one given here it does not necessarily mean that solution is incorrect. Exploring and making sense of such differences can be an excellent way to deepen one's understanding of the material, and I encourage the sharing of alternate solutions with classmates and/or the instructor.

I also encourage the use of this manual as a tool for getting unstuck on a problem, but to go no further in the solution than is necessary to get past the sticking point. It can be tempting to turn to the solution and read through it in its entirety; however, to get the most out of this manual I suggest working through the solution only far enough to see how to continue. Finally, at some point, especially during exam preparation, it is important to make sure that all assigned problems can be worked from scratch without assistance. Reviewing solutions is valuable, but a much deeper level of understanding is required to reproduce solutions independently, and that is what is usually required on an exam. Exercises in this manual are referenced by section number. Thus "Exercise 1.2.5" refers to Exercise 5 from Section 1.2.

JEFFREY T. BARTON
December 2015

ABOUT THE COMPANION WEBSITE

This book is accompanied by a companion website:
www.wiley.com/go/barton/solutionsmanual_modelsforlife

The website includes:

- Instructor's Solutions Manual
- Instructor Excel Spreadsheets for Models in Text
- Test Bank
- Instructor Notes and Project Ideas
- Student Excel Spreadsheets
- Selected Excel Spreadsheets Used in Exercises for Instructor

1

DENSITY INDEPENDENT POPULATION MODELS

1.1 EXPONENTIAL GROWTH

1 Consider the flow diagram in Text Figure 1.21.

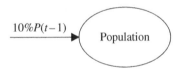

$10\%P(t-1)$ → Population

TEXT FIGURE 1.21 Flow diagram for Exercise 1.1.1.

a. Find the corresponding DDS.

The flow diagram tells us how the population changes from one year to the next. Inward pointing arrows represent additions while outward pointing arrows represent subtractions. Here there is only one arrow, and it represents an addition. Thus the DDS is given by

$$P(t) = P(t-1) + 10\%P(t-1).$$

We can also write the DDS as $P(t) = P(t-1) + 0.10P(t-1)$, or $P(t) = 1.10P(t-1)$.

Solutions Manual to Accompany Models for Life: An Introduction to Discrete Mathematical Modeling with Microsoft® Office Excel®, First Edition. Jeffrey T. Barton.
© 2016 John Wiley & Sons, Inc. Published 2016 by John Wiley & Sons, Inc.
Companion website: www.wiley.com/go/barton/solutionsmanual_modelsforlife

b. Use a calculator to predict the population after 2 years if $P(0) = 50$.

If $P(0) = 50$, then by using the DDS we can predict the population 1 year later:

$$P(1) = P(0) + 10\% P(0)$$
$$= 50 + 0.10 \cdot 50$$
$$= 55.$$

Applying the DDS once more gives us the model prediction for year 2:

$$P(2) = P(1) + 10\% P(1)$$
$$= 55 + 0.10 \cdot 55$$
$$= 60.5.$$

After 2 years we predict the population will be 60.5.

c. Use Excel to project the population in year 10.

Since the model in this problem is the exponential growth model, we can save time by using the same spreadsheet we created for the Yellowstone grizzly population. We only need to change the growth rate to $r = 10\%$ and the initial population to $P(0) = 50$. Figure 1.1 shows the result with the projection for year 10 highlighted. The model predicts a population of about 129.7 in year 10.

	A	B	C
1	Exercise 1.1.1		
2			
3	Growth rate, $r=$		10%
4			
5	t	Population	
6	0	50	
7	1	55.0	
8	2	60.5	
9	3	66.6	
10	4	73.2	
11	5	80.5	
12	6	88.6	
13	7	97.4	
14	8	107.2	
15	9	117.9	
16	10	129.7	

FIGURE 1.1   Excel output for Exercise 1.1.1.

3 Consider the flow diagram in Text Figure 1.23.

TEXT FIGURE 1.23 Flow diagram for Exercise 1.1.3.

a. Find the corresponding DDS.

The flow diagram tells us how the population changes from one year to the next. Inward pointing arrows represent additions while outward pointing arrows represent subtractions. Here we have two arrows: one an addition and one a subtraction. Thus the DDS is given by

$$P(t) = P(t-1) + 8\%P(t-1) - 5\%P(t-1)$$
$$= P(t-1) + 3\%P(t-1).$$

We can also write the DDS as $P(t) = P(t-1) + 0.03P(t-1)$, or $P(t) = 1.03P(t-1)$.

b. Use a calculator to predict the population after 2 years if $P(0) = 100$.

If $P(0) = 100$, then by using the DDS we can predict the population 1 year later:

$$P(1) = P(0) + 3\%P(0)$$
$$= 100 + 0.03 \cdot 100$$
$$= 103.$$

Applying the DDS once more gives us the model prediction for year 2:

$$P(2) = P(1) + 3\%P(1)$$
$$= 103 + 0.03 \cdot 103$$
$$= 106.09.$$

After 2 years we predict the population will be about 106.1.

c. Use Excel to project the population in year 10.

We see from the DDS that this model is still an exponential growth model with $r = 3\%$. Thus we can use the Yellowstone grizzly spreadsheet with the new growth rate and the initial population set to 100. The result is given in Figure 1.2 with the projection for year 10 highlighted. The model predicts a population of about 134.4 in year 10.

	A	B	C
1	Exercise 1.1.3		
2			
3	Growth rate, $r=$		3%
4			
5	t	Population	
6	0	100	
7	1	103.0	
8	2	106.1	
9	3	109.3	
10	4	112.6	
11	5	115.9	
12	6	119.4	
13	7	123.0	
14	8	126.7	
15	9	130.5	
16	10	134.4	

FIGURE 1.2   Excel output for Exercise 1.1.3.

5   Draw a flow diagram that corresponds to the following DDS:

$$P(t) = P(t-1) + 4\% P(t-1).$$

The addition of 4% of the previous year's population is represented by an inward pointing arrow in the flow diagram, given in Figure 1.3.

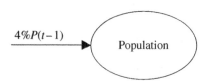

$4\% P(t-1)$ Population

FIGURE 1.3   Flow diagram for Exercise 1.1.5.

7   Draw a flow diagram that corresponds to the following DDS:

$$P(t) = P(t-1) - 0.30 P(t-1).$$

The DDS indicates a subtraction of 30% of the previous year's population. We account for this subtraction with an outward pointing arrow in the flow diagram, given in Figure 1.4. Note that there is no minus sign in front of the arrow label.

FIGURE 1.4 Flow diagram for Exercise 1.1.7.

9 Give the flow diagram and corresponding DDS for a grizzly population that is growing by 8% per year and has 5 bears illegally poached annually.

We represent the 8% growth by an inward pointing arrow and the poaching by an outward pointing arrow. The result is Figure 1.5. Note that there is no minus sign in front of the 5.

FIGURE 1.5 Flow diagram for Exercise 1.1.9.

The corresponding DDS is given by $P(t) = P(t-1) + 0.08 \cdot P(t-1) - 5$.

11 Suppose you know that the DDS for a population is given by

$$P(t) = P(t-1) + 3\%P(t-1) - 50.$$

a. Draw a flow diagram that would lead to this DDS.

The 3% increase is represented by an inward pointing arrow while the removal of 50 from the population is represented by an outward pointing arrow. The result is given in Figure 1.6.

FIGURE 1.6 Flow diagram for Exercise 1.1.11.

b. Explain in a complete sentence how the population is changing from year to year.

The population is experiencing growth of 3% of the previous year's population while at the same time 50 members of the population are leaving each year.

13   Suppose that the *1993 Grizzly Bear Recovery Plan* had never been implemented
        and that the 1993 estimate of a 1% growth rate continued to hold. How long
        would it have taken for the population to reach 416 bears?

        We use the Yellowstone grizzly population Excel model with $r = 1\%$
        and $P(0) = 197$. We are looking for the year that the population reaches 416
        bears, so we drag the model formulas down until we see the population meet
        or exceed 416 for the first time. This happens 76 years from the initial popula-
        tion estimate, and the population of bears is projected to be about 419.7 at
        that time.

15   Suppose that the numbers of adult females with cubs sighted in Yellowstone
        were 52 in 2003, 60 in 2004, and 65 in 2005. Estimate the total grizzly popu-
        lation in 2005.

        The 3-year total of adult female grizzlies is $52 + 60 + 65 = 177$. No known
        deaths are mentioned, so we assume 0 known deaths. Thus we have 177 adult
        females, representing about 27.4% of the total population of bears. This total is
        given by $\frac{177}{0.274} \approx 645.99$, or about 646 bears.

17   Text Table 1.2 contains more population data for the wild California condor
        population from the 1996 *Recovery Plan for the California Condor* (U.S. Fish
        and Wildlife Service, 1996).

        **TEXT TABLE 1.2   The Number of California Condors Remaining in
        the Wild between 1982 and 1985 (U.S. Fish and Wildlife Service, 1996)**

Year	Number Wild California Condors
1982	21
1983	19
1984	15
1985	9

   a. Compare the population values in the table to what our model would predict
        using the rate of decline found in Example 1.5 and an initial population of
        50 condors. In general, how well did our model do?

        Here we use the California condor Excel spreadsheet that we already created,
        where $P(0) = 50$ in 1968, and the rate of decline from Example 1.5 is
        $r = 6.6\%$. Next we drag the model formulas down until we reach the year
        1985, or $t = 17$. The projected values for years 1982-5 for our model are
        19, 18, 17, and 16. We compare the model projections to the data in Text
        Table 1.2, which recorded condor populations of 21, 19, 15, and 9 for the
        years 1982-5. Our model seems to have done reasonably well, though from
        the data it appears as though something happened in 1985 that caused a lar-
        ger than predicted decline in the population.

b. Can you think of possible reasons for any discrepancies?

As noted above, the most striking difference between our model projections and the actual population data seems to be for the year 1985. There could be any number of reasons for the larger than predicted decline in 1985, including accidents, poaching, or disease.

19 Recall that our estimate for the California condor's rate of decline was based on the lower population estimates given by Sibley, Mailed, and Wilbur. Re-estimate the rate of decline from 1968 to 1978 using three other combinations from the population estimates:

a. The lower value from 1960's and the higher value from 1978.

The range of values for the California condor population was given as 50–60 in the late 1960's and 25–30 in 1978. Taking the lower value from the 1960's (with the assumption of 1968 for our starting year), we use $P(0) = 50$. Using the higher estimate, 30 condors, in 1978 gives us $P(10) = 30$. Thus we repeat the trial-and-error approach from Example 1.5 in order to estimate the rate of decline from 1968 to 1978. We use the already created California condor Excel model and type in different values for r until we get 30 condors in 1978. The result is shown in Figure 1.7 with the value for r highlighted. Our new estimate for the rate of decline is about 5% per year. Note that it makes sense for the rate of decline to be lower than in Example 1.5 because the assumed population in 1978 is higher – there was less of an assumed decline.

b. The higher value from 1960's and the lower value from 1978.

	A	B	C
1	Exercise 1.19		
2			
3	Rate of decline, $r=$		5.0%
4			
5	t	Population	
6	0	50	
7	1	48	
8	2	45	
9	3	43	
10	4	41	
11	5	39	
12	6	37	
13	7	35	
14	8	33	
15	9	32	
16	10	30	

FIGURE 1.7   Excel output for Exercise 1.1.19.

Here we need to use $P(0) = 60$ and $P(10) = 25$. Repeating the trial-and-error exercise from part a. gives us the estimate $r = 8.4\%$. Note that it makes sense for the rate of decline to be higher than in Example 1.5 because the assumed population in 1968 is higher – there is more of an assumed decline to 1978.

c. The higher value from 1960's and the higher value from 1978.

Here we need to use $P(0) = 60$ and $P(10) = 30$. Repeating the trial-and-error exercise from part a. gives us the estimate $r = 6.6\%$. Note that it makes sense for the rate of decline to be the same as in Example 1.5 because the assumed population declines by 50% from 1968 to 1978, just as it did in Example 1.5 when the values used were 50 in 1968 and 25 in 1978.

d. How much difference do you see in r?

The range of values for r is $5\% - 8.4\%$, depending on the values used from the given data estimates. The value we used in the text, $r = 6.6\%$, falls squarely in that range and thus seems a reasonable choice.

1.2   EXPONENTIAL GROWTH WITH STOCKING OR HARVESTING

1   Consider the flow diagram in Text Figure 1.33.

a. Find the corresponding DDS.

TEXT FIGURE 1.33   Flow diagram for Exercise 1.2.1.

Recalling that inward pointing arrows represent additions and outward pointing arrows subtractions, the DDS is given by $P(t) = P(t-1) + 0.08P(t-1) - 50$.

b. Use a calculator to predict the population after 2 years if $P(0) = 650$.

According to the DDS we find the population after 1 year by calculating

$$P(1) = P(0) + 0.08P(0) - 50$$
$$= 650 + 0.08 \cdot 650 - 50$$
$$= 652.$$

Applying the DDS once more give us our projection for year 2:

$$P(2) = P(1) + 0.08P(1) - 50$$

$$= 652 + 0.08 \cdot 652 - 50$$

$$= 654.16.$$

Thus after 2 years our model predicts a population of about 654.

c. Use Excel to project the population after 15 years.

The model in this problem is in the form of an exponential growth model with harvesting. Thus we can use the white-tailed deer Excel model, being careful to enter the appropriate parameter values. Once we have entered the correct parameter values we drag the model formulas down to year 15 and record the result. The result is a population of about 704 as shown in Figure 1.8.

	A	B	C
1	Exercise 1.2.1		
2			
3	Growth rate, r =		8.0%
4	Harvest num., a =		50
5			
6	t	Population	
7	0	650	
8	1	652.0	
9	2	654.2	
10	3	656.5	
11	4	659.0	
12	5	661.7	
13	6	664.7	
14	7	667.8	
15	8	671.3	
16	9	675.0	
17	10	679.0	
18	11	683.3	
19	12	688.0	
20	13	693.0	
21	14	698.4	
22	15	704.3	

FIGURE 1.8   Excel output for Exercise 1.2.1.

3   Draw a flow diagram that corresponds to the following DDS:

$$P(t) = P(t-1) + 4\%P(t-1) - 40.$$

The addition of 4% of the previous year's population is represented by an inward pointing arrow while the subtraction of 40 is represented by an outward pointing arrow. The result is given in Figure 1.9. Note that there is no minus sign in front of the "40."

FIGURE 1.9   Flow diagram for Exercise 1.2.3.

5 *Extension*: In Example 1.8 we determined that an average of 12.23 Mississippi sandhill cranes were captive-reared and released annually between 1981 and 1993. Text Table 1.3 gives the actual numbers, taken from Table 2 in Valentine and Lohoefener (1991), for the years 1981–1990.

TEXT TABLE 1.3   The Number of Mississippi Sandhill Cranes Captive-Reared and Released between 1981 and 1990 (Valentine & Lohoefner, 1991)

Year of Release	Number Captive-released
1981	9
1982	4
1983	8
1984	4
1985	10
1986	7
1987	2
1988	10
1989	13
1990	29

a. How does 12.23 compare to the average number of cranes that were actually released?

Here we find the average of the values given in Text Table 1.3:

$$\frac{(9+4+8+4+10+7+2+10+13+29)}{10} = 9.6.$$

Thus the actual average number of cranes released from 1981–90 was 9.6 – about 2.6 less than the average number we estimated for 1981–93.

b. What factor(s) might account for the difference?

One possibility is that the table of data only extends from 1981 to 1990, so there could have been large cohorts of cranes released in 1991, 1992, and 1993. It is also possible that the assumed rate of decline used in Example 1.8 (6%) turned out to be higher than what the population actually experienced, thus causing us to overestimate the number of cranes released.

c. Using a rate of decline of 6%, and the actual release values from 1981 to 1990, estimate how many cranes there were in 1990. (Note: this exercise will require a significant modification of the crane Excel spreadsheet.)

The difference in this problem versus what we have done previously is that there is no longer a single stocking number that we can refer to for all years. Instead, we have a different stocking number each year that must be added. We can arrange this by including a separate column in our Excel model where we store the release data, and then referring to that column for the stocking number each year. The set-up is given in Figure 1.10. Note how the population formula now refers to column C for the stocking number each year. Figure 1.11 shows the numerical result. The model predicts about 107 cranes in 1990.

	A	B	C
1	Exercise 1.2.5		
2			
3	Rate of decline, r =		0.06
4	Stocking num., a =		12.23
5			
6	t	Population	Release Data
7	0	50	
8	=A7+1	=B7-$C$3*B7+C8	9
9	=A8+1	=B8-$C$3*B8+C9	4
10	=A9+1	=B9-$C$3*B9+C10	8
11	=A10+1	=B10-$C$3*B10+C11	4
12	=A11+1	=B11-$C$3*B11+C12	10
13	=A12+1	=B12-$C$3*B12+C13	7
14	=A13+1	=B13-$C$3*B13+C14	2
15	=A14+1	=B14-$C$3*B14+C15	10
16	=A15+1	=B15-$C$3*B15+C16	13
17	=A16+1	=B16-$C$3*B16+C17	29

FIGURE 1.10   Excel set-up for Exercise 1.2.5.

d. Determine, on average, how many cranes must have been released in 1991, 1992, and 1993 in order to end up with 135 cranes in 1993.

After the year 1990 we no longer have release data, so from that year on we refer to the stocking parameter that we kept in cell C4. The formula version of the spreadsheet is given in Figure 1.12. Our task is to experiment with different stocking numbers until we end up with 135 cranes in 1993. The result is a stocking number of about 16.3 cranes per year.

	A	B	C
1	Exercise 1.2.5		
2			
3	Rate of decline, r =		6.0%
4	Stocking num., a =		12.23
5			
6	t	Population	Release Data
7	0	50	
8	1	56.0	9
9	2	56.6	4
10	3	61.2	8
11	4	61.6	4
12	5	67.9	10
13	6	70.8	7
14	7	68.6	2
15	8	74.4	10
16	9	83.0	13
17	10	107.0	29

FIGURE 1.11   Excel output for Exercise 1.2.5.

	A	B	C
1	Exercise 1.2.5		
2			
3	Rate of decline, r =		0.06
4	Stocking num., a =		12.23
5			
6	t	Population	Release Data
7	0	50	
8	=A7+1	=B7-$C$3*B7+C8	9
9	=A8+1	=B8-$C$3*B8+C9	4
10	=A9+1	=B9-$C$3*B9+C10	8
11	=A10+1	=B10-$C$3*B10+C11	4
12	=A11+1	=B11-$C$3*B11+C12	10
13	=A12+1	=B12-$C$3*B12+C13	7
14	=A13+1	=B13-$C$3*B13+C14	2
15	=A14+1	=B14-$C$3*B14+C15	10
16	=A15+1	=B15-$C$3*B15+C16	13
17	=A16+1	=B16-$C$3*B16+C17	29
18	=A17+1	=B17-$C$3*B17+$C$4	
19	=A18+1	=B18-$C$3*B18+$C$4	
20	=A19+1	=B19-$C$3*B19+$C$4	

FIGURE 1.12   Excel set-up for Exercise 1.2.5d.

7 When deciding on a real harvesting strategy, it is not just the total number harvested that is important. Rather, the sex ratio of the harvest number is also very important. In fact, Palmer and Storm write, "During the past decade, deer populations in the Northeast have continued to increase except in states that harvested markedly more antlerless than antlered deer. (Palmer & Storm, 1995)" In a sentence or two, discuss why the sex ratio of deer harvests should play such an important role in controlling deer populations.

Imagine two populations of deer: each has 1,000,000 male deer and 1,000,000 female deer for a total of 2,000,000 in each population. Suppose that from Population 900,000 male deer are harvested, while from Population 900,000 female deer are harvested. We would expect that even though the same number of deer were harvested from each, Population 1 will experience much more growth in the following year than Population 2 because all of the female deer still have the potential to produce offspring.

1.3   TWO FUNDAMENTAL EXCEL TECHNIQUES

1 Consider the DDS below:

$$P(t) = P(t-1) + 0.10P(t-1) - 50.$$

a. Graph the population over a period of 10 years if $P(0) = 550$.

This is a straightforward graphing exercise where we have a model representing an exponentially growing population that is undergoing harvesting. We copy the formula down to year 10, select both the time and population columns including the column headings, and insert an X,Y-Scatter graph. The result is shown in Figure 1.13.

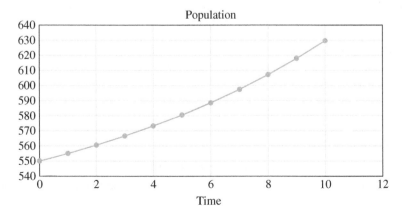

FIGURE 1.13   Excel graph for Exercise 1.3.1a.

b. Graph the population over a period of 10 years if $P(0) = 450$.

Here we need only change the initial population from 550 to 450, and Excel will automatically update our graph to give the one shown in Figure 1.14.

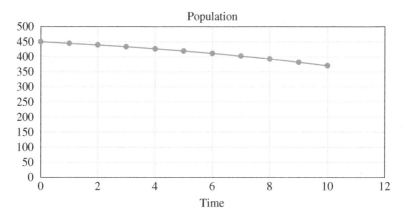

FIGURE 1.14   Excel graph for Exercise 1.3.1b.

c. Describe the difference in the behavior of the population in the two cases.

In the first case we see an increasing population while in the second case the population is decreasing. With an initial population of 450 the 10% growth rate is not enough to overcome the effects of a harvesting level of 50.

3 Describe the difference in population behavior between Exercises 1.3.1 and 1.3.2. In Exercise 1.3.1, the first population increases without bound while the second population crashes to extinction. Neither of these behaviors is present in Exercise 1.3.2. In Exercise 1.3.2, the decreasing population decreases less and less over time, while the increasing population increases less and less over time. Both of the populations in Exercise 1.3.2 will eventually level off, or stabilize, at a positive population value, and we can find that value by dragging the model formula down far enough that the population is no longer changing.

5 Sketch a graph by hand for each of the following situations.
   a. A population where $P(0) = 100$ and the population is increasing over time at an increasing rate.

   An example sketch is provided in Figure 1.15.

   b. A population where $P(0) = 200$ and the population is increasing over time at a decreasing rate.

   An example sketch is provided in Figure 1.16.

   c. A population where $P(0) = 50$ and the population is decreasing over time at a decreasing rate.

   An example sketch is provided in Figure 1.17.

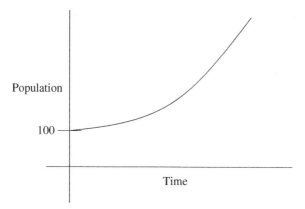

FIGURE 1.15   Example graph for Exercise 1.3.5a.

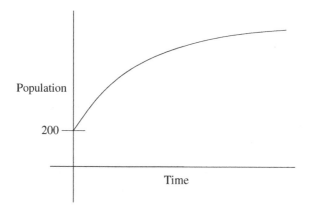

FIGURE 1.16   Example graph for Exercise 1.3.5b.

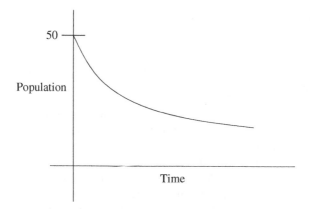

FIGURE 1.17   Example graph for Exercise 1.3.5c.

d. A population where $P(0) = 100$ and the population is decreasing over time at an increasing rate.

An example sketch is provided in Figure 1.18.

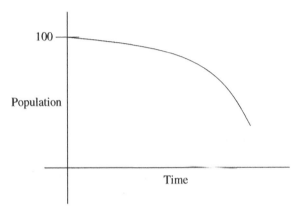

FIGURE 1.18   Example graph for Exercise 1.3.5d.

7 Consider the DDS given by $P(t) = P(t-1) + rP(t-1) - 50$. If the initial population is 400, use Goal Seek to determine the value for r that results in a population of 800 10 years later.

   Here we use the Excel model for a population that is growing exponentially and undergoing harvesting. Since the growth rate is unknown, we use a stand-in value of $r = 5\%$ initially. We also set the harvesting number to 50 and the initial population to 400. We need to copy the formula down to year 10 and then use Goal Seek to find the growth rate that produces a population of 800 in year 10. The set-up just before Goal Seek is run is shown in Figure 1.19. The result is a growth rate of about 17%.

9 Regarding white-tailed deer, recall the Curtis and Sullivan (Curtis & Sullivan, 2001) estimate that deer populations can double every 2–3 years. In Examples 1.9 and 1.10 we based our Excel work on a doubling time of 3 years.

   a. Use Goal Seek to rework Examples 1.9 and 1.10 in the text, this time assuming a doubling time of 2 years.

   Reworking Example 1.9 means we need to find the growth rate, r, that would result in a doubling of the deer population in 2 years. We can start with any initial deer population, and we use our deer Excel model to find the correct r (making sure that the harvest number is set to 0 before proceeding). The set-up just before Goal Seek is run is given in Figure 1.20. The result is an annual growth rate of approximately 41.4%.

   Reworking Example 1.10 can also be with Goal Seek. The initial number of deer is 3,000,000, and we use Goal Seek to find the new harvesting number that keeps

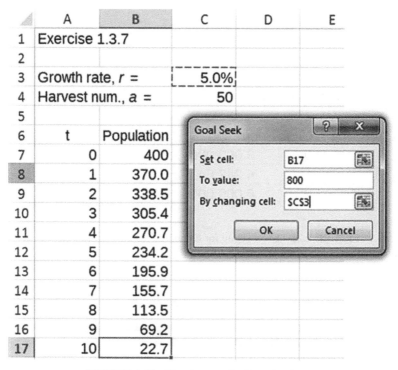

FIGURE 1.19   Excel set-up for Exercise 1.3.7.

FIGURE 1.20   Excel set-up for Example 1.9 in Exercise 1.3.9.

the population from growing. The set-up just before Goal Seek is run is given in Figure 1.21. The resulting harvesting number is about 1,242,641 deer per year.

b. How does the new harvesting number compare to the previous estimate of 780,000?

The harvesting number turns out to be higher than the original 780,000.

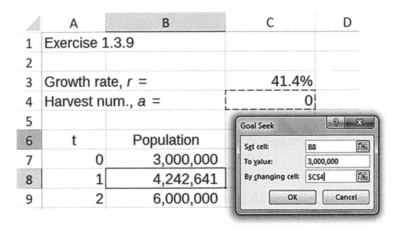

FIGURE 1.21   Excel set-up for Example 1.10 in Exercise 1.3.9.

c. Explain why the answer for b. makes sense in the context of the problem.

Because we assume a shorter doubling time and get an increased growth rate as a result, we should expect to have to harvest more deer to keep the population from growing.

1.4   EXPLICIT FORMULAS

For all of the exercises below, use the appropriate explicit formula to find the solution.

1   Consider the DDS given by $P(t) = P(t-1) + 0.10P(t-1)$. Determine the population in year 5 if the initial population is 400.

We use the explicit formula for exponential growth and plug in the relevant parameters. With $t = 5$, $P(0) = 400$, and $r = 0.10$, we have

$$P(t) = (1+r)^t P(0)$$
$$P(5) = (1+.10)^5 400$$
$$= 644.204.$$

The population in year 5 is projected to be about 644.

3   Consider the DDS given by $P(t) = P(t-1) + 0.10P(t-1) - 5$. Determine the population in year 5 if the initial population is 400.

We use the explicit formula for exponential growth with harvesting and plug in the relevant parameters. With $t = 5$, $P(0) = 400$, $r = 0.10$, and $a = -5$, we have

$$P(t) = (1+r)^t P(0) + a\frac{(1+r)^t - 1}{r}$$

$$P(5) = (1+.10)^5 400 - 5\frac{(1+.10)^5 - 1}{.10}$$

$$= 644.204 - 30.5255$$

$$\approx 613.68.$$

The population in year 5 is projected to be about 614.

5 Consider the DDS given by $P(t) = P(t-1) + 0.10P(t-1) - a$. If the initial population is 500, determine the value for a that results in a population of 600 12 years later.

Here we use the explicit formula for exponential growth with harvesting, and we have to solve for the unknown a. Plugging in all known parameters gives

$$P(t) = (1+r)^t P(0) + a\frac{(1+r)^t - 1}{r}$$

$$P(12) = (1+.10)^{12} 500 - a\frac{(1+.10)^{12} - 1}{.10}.$$

We have introduced a minus sign in front of the a in order to maintain consistency with the problem statement. Next we note that if the population in year 12 is to be 600 then we know $P(12) = 600$. Thus we have

$$600 = (1+.10)^{12} 500 - a\frac{(1+.10)^{12} - 1}{.10}$$

$$600 \approx 1569.21 - a \cdot 21.38$$

$$21.38a \approx 969.21$$

$$a \approx \frac{969.21}{21.38} \approx 45.33.$$

The harvesting number would need to be about 45.33 per year in order for the population to be 600 in year 12.

7 Given the initial population estimate of 197 Yellowstone grizzlies in 1993 and the later estimate of 416 Yellowstone grizzlies in 2002, we found that the population grew by about 8.65% per year.

   a. Using the 8.65% growth rate, what would your model predict for the population in the year 2193?

   Here we use the explicit formula for exponential growth with $r = 0.0865$, $P(0) = 197$, and $t = 200$:

$$P(t) = (1+r)^t P(0)$$

$$P(200) = (1.0865)^{200} 197$$

$$P(200) = 3,165,281,631.27.$$

The population in the year 2193 is predicted to be about 3.2 billion bears.

b. Does your answer in part a. seem reasonable? Why or why not?

No, 3.2 billion is unreasonably large for the Yellowstone grizzly population.

c. Suppose that the 1993 Grizzly Bear Recovery Plan had never been implemented and that the 1993 estimate of a 1% growth rate continued to hold. How long would it have taken for the population to reach 416 bears?

Here we use the explicit formula for exponential growth where we know the values of all parameters except time, t. With $r = 0.01$, $P(0) = 197$, and a goal population of 416 bears, we have

$$P(t) = (1+r)^t P(0)$$

$$416 = (1.01)^t 197$$

$$2.112 \approx (1.01)^t.$$

Here we can use trial and error with a calculator to find t, or we can take the natural logarithm of both sides, which has the effect of bringing any exponents down in front of the logarithm. We get

$$\ln(2.112) \approx \ln(1.01^t)$$

$$0.7476 \approx t \cdot \ln(1.01)$$

$$0.7476 \approx t \cdot 0.00995$$

$$\frac{0.7476}{0.00995} \approx t$$

$$75.14 \approx t.$$

It would have taken just over 75 years for the grizzly population to reach 416 bears had the growth rate remained at 1% each year.

9 Consider the explicit formula for our harvesting/stocking model. Show that if there is no stocking or harvesting, then the formula is the same as the explicit formula for plain exponential growth.

Here we begin with the explicit formula for exponential growth with harvesting or stocking, and we plug in 0 for the harvesting/stocking number:

$$P(t) = (1+r)^t P(0) + a\frac{(1+r)^t - 1}{r}$$

$$P(t) = (1+r)^t P(0) + 0\frac{(1+r)^t - 1}{r}$$

$$P(t) = (1+r)^t P(0).$$

The result is the explicit formula for exponential growth.

11 The U.S. Census in 2000 (see www.census.gov) estimated the population of the United States to be 281.4 million. Without immigration, the population would grow by approximately 0.6% each year. Data available at www.census.gov indicates that approximately 1,000,000 immigrants enter the U.S. each year.

a. Create a flow diagram for the U.S. population.

A flow diagram for the U.S. population is given in Figure 1.22.

FIGURE 1.22 Flow diagram for Exercise 1.4.11.

b. From the flow diagram, give the DDS.

Each inward pointing arrow represents an addition to the previous year's population, so the DDS is given by

$$P(t) = P(t-1) + 0.006 \cdot P(t-1) + 1,000,000.$$

c. Use the explicit formula to predict the U.S. population in the year 2050.

With 281,400,000 as our initial population in the year 2000, $t = 50$. Thus we have

$$P(50) = (1.006)^{50} 281,400,000 + 1,000,000\frac{(1.006)^{50} - 1}{.006}$$

$$P(50) \approx 379,509,917 + 58,108,219$$

$$P(50) \approx 437,618,136.$$

Under the assumed conditions the U.S. population will be approximately 437.6 million in the year 2050.

d. How does your projection compare to the 419.9 million projected for 2050 by the U.S. Census?

Our projection is about 4% higher than the U.S. Census projection. This indicates that either our assumed growth rate is too high for the years 2000–2050, our assumed immigration rate is too high, or both.

13 *Extension*: Following the spirit of the derivation for the explicit formula for an affine model, find the explicit formula for the general model described in Exercise 1.4.12. Confirm your result in 1.4.12d by using the explicit formula.

The DDS for the general model in Exercise 1.4.12 is given by

$$P(t) = P(t-1) + rP(t-1) + a_0(1+s)^{t-1}$$
$$= (1+r)P(t-1) + a_0(1+s)^{t-1},$$

where r, a_0, and s are all parameters that can be positive or negative. Our task is to work step-by-step through a few iterations of the DDS until we can see a pattern.

For $t=1$ we get

$$P(1) = (1+r)P(0) + a_0(1+s)^{1-1}$$
$$= (1+r)P(0) + a_0(1+s)^0$$
$$= (1+r)P(0) + a_0.$$

For $t=2$ we get

$$P(2) = (1+r)P(1) + a_0(1+s)^{2-1}$$
$$= (1+r)P(1) + a_0(1+s).$$

After substituting for $P(1)$ we get

$$P(2) = (1+r)[(1+r)P(0) + a_0] + a_0(1+s)$$
$$= (1+r)^2 P(0) + (1+r)a_0 + a_0(1+s).$$

For $t=3$ we get

$$P(3) = (1+r)P(2) + a_0(1+s)^{3-1}$$
$$= (1+r)P(2) + a_0(1+s)^2.$$

After substituting for $P(2)$ we get

$$P(3) = (1+r)\left[(1+r)^2 P(0) + (1+r)a_0 + a_0(1+s)\right] + a_0(1+s)^2$$
$$= (1+r)^3 P(0) + (1+r)^2 a_0 + a_0(1+r)(1+s) + a_0(1+s)^2.$$

For $t=4$ we get

$$P(4) = (1+r)P(3) + a_0(1+s)^{4-1}$$
$$= (1+r)P(3) + a_0(1+s)^3.$$

After substituting for $P(3)$ we get

$$P(4) = (1+r)\left[(1+r)^3 P(0) + (1+r)^2 a_0 + a_0(1+r)(1+s) + a_0(1+s)^2\right] + a_0(1+s)^3$$
$$= (1+r)^4 P(0) + (1+r)^3 a_0 + a_0(1+r)^2(1+s) + a_0(1+r)(1+s)^2 + a_0(1+s)^3.$$

The pattern seems to be that for any t we have

$$P(t) = (1+r)^t P(0) + a_0\left[(1+r)^{t-1} + (1+r)^{t-2}(1+s) + \cdots + (1+r)(1+s)^{t-2} + (1+s)^{t-1}\right].$$

Next we need to do some algebra on the sum inside the brackets. For starters, we factor out the term $(1+r)^{t-1}$ to get

$$P(t) = (1+r)^t P(0) + a_0(1+r)^{t-1}\left[1 + (1+r)^{-1}(1+s) + \cdots + (1+r)^{-t+2}(1+s)^{t-2}\right.$$
$$\left. + (1+r)^{-t+1}(1+s)^{t-1}\right].$$

Placing terms with negative exponents in denominators gives

$$P(t) = (1+r)^t P(0) + a_0(1+r)^{t-1}\left[1 + \frac{(1+s)}{(1+r)} + \cdots + \frac{(1+s)^{t-2}}{(1+r)^{t-2}} + \frac{(1+s)^{t-1}}{(1+r)^{t-1}}\right].$$

Then we have

$$P(t) = (1+r)^t P(0) + a_0(1+r)^{t-1}\left[1 + \left(\frac{1+s}{1+r}\right) + \cdots + \left(\frac{1+s}{1+r}\right)^{t-2} + \left(\frac{1+s}{1+r}\right)^{t-1}\right].$$

We can now recognize the sum inside the brackets as a geometric series with ratio $x = \frac{1+s}{1+r}$. This observation allows us to rewrite the sum using the geometric series formula to get

$$P(t) = (1+r)^t P(0) + a_0(1+r)^{t-1}\left[\frac{\left(\frac{1+s}{1+r}\right)^t - 1}{\left(\frac{1+s}{1+r}\right) - 1}\right].$$

We can use the formula as is, or we can employ some further simplification with exponents to yield the more succinct version

$$P(t) = (1+r)^t P(0) + a_0 \left[\frac{(1+s)^t - (1+r)^t}{s-r} \right],$$

The last task is to confirm the result of Exercise 1.4.12d by plugging in all required parameters. With $r = -0.06$, $a_0 = 100$, $s = 0.10$, $P(0) = 400$, and $t = 10$, we get

$$P(t) = (1+r)^t P(0) + a_0 \left[\frac{(1+s)^t - (1+r)^t}{s-r} \right]$$

$$P(10) = (.94)^{10} 400 + 100 \left[\frac{(1.10)^{10} - (.94)^{10}}{0.10 - (-.06)} \right]$$

$$\approx 215.45 + 1284.45$$

$$\approx 1499.9.$$

The result is about 1500 cranes after 10 years, and this agrees with the Excel result from 12d.

1.5   EQUILIBRIUM VALUES AND STABILITY

1   Consider the DDS given by $P(t) = P(t-1) + .05P(t-1) - 10$.

    a. Find all equilibrium values for the DDS.

    We need to find P^* such that

$$P^* = P^* + .05P^* - 10.$$

    We have

$$0 = .05P^* - 10$$
$$10 = .05P^*$$
$$200 = P^*.$$

    Thus the only equilibrium value is $P^* = 200$.

    b. Use Excel to confirm that the values found in a. are in fact equilibrium values.

    To confirm that $P^* = 200$ is an equilibrium value in Excel we need to enter the DDS and verify that if the DDS starts at 200, it stays at 200. Figure 1.23 shows the formula for the DDS along with the verification we need.

	A	B
1	Exercise 1.5.1	
2		
3	t	P(t)
4	0	200
5	1	=B4+0.05*B4-10
6	2	200
7	3	200
8	4	200
9	5	200
10	6	200
11	7	200
12	8	200

FIGURE 1.23   Excel confirmation for Exercise 1.5.1.

c. Determine the stability of any equilibrium values found in a. by producing an appropriate Excel graph.

We graph the DDS for different starting points near the equilibrium value all on the same axes. Figure 1.24 shows the result. Because populations that start off of the equilibrium value continue to get farther away from it, the equilibrium at 200 is unstable. We also note that the horizontal line at 200 provides graphical confirmation that it is in fact an equilibrium value.

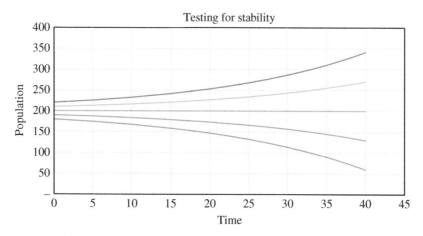

FIGURE 1.24   Excel graph testing for stability in Exercise 1.5.1.

3   *Extension*: Consider the DDS given by $P(t) = P(t-1) + 0.004(5 - P(t-1)) \cdot P(t-1)$.

a. Find all equilibrium values for the DDS.

We need to find all P^* such that

$$P^* = P^* + 0.004(5 - P^*) \cdot P^*.$$

We have

$$0 = 0.004(5 - P^*) \cdot P^*.$$

A product of real numbers is only zero if at least one of the factors is zero, so we must have either $P^* = 0$ or $5 - P^* = 0$. Thus our two equilibrium values are $P^* = 0$ and $P^* = 5$.

b. Use Excel to confirm that the values found in a. are in fact equilibrium values.

First we need to implement the DDS in Excel. Figure 1.25 shows the correct Excel formula. To verify that 0 and 5 are equilibrium values, we plug those values in for the initial population and note that the population does not change from those values. Figure 1.26 shows the verification for the equilibrium value at 5.

	A	B
1	Exercise 1.5.3	
2		
3	t	P(t)
4	0	100.00
5	1	=B4+0.004*(5-B4)*B4
6	2	47.86

FIGURE 1.25   Excel set-up for Exercise 1.5.3.

	A	B
1	Exercise 1.5.3	
2		
3	t	P(t)
4	0	5.00
5	1	5.00
6	2	5.00
7	3	5.00
8	4	5.00
9	5	5.00
10	6	5.00
11	7	5.00
12	8	5.00

FIGURE 1.26   Excel confirmation of equilibrium value for Exercise 1.5.3.

c. Determine the stability of any equilibrium values found in a. by producing an appropriate Excel graph.

For each equilibrium value we graph the DDS for different nearby starting points all on the same axes. Figure 1.27 shows the result. Populations that start a little above 0 are moving away from 0, but populations that start off of 5 are moving toward 5. Thus we say that the equilibrium at 0 is unstable and that the equilibrium at 5 is stable. We also note that the horizontal line at 5 provides graphical confirmation that it is in fact an equilibrium value.

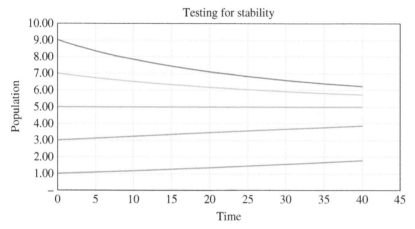

FIGURE 1.27   Excel graph testing for stability in Exercise 1.5.3.

5   *Extension*: Consider the DDS given by

$$P(t) = P(t-1) + 0.05 \left(1 - \frac{P(t-1)}{10,000} \right) P(t-1) - 125.$$

a. Find all equilibrium values for the DDS.

We need to find all P^* such that

$$P^* = P^* + 0.05 \left(1 - \frac{P^*}{10,000} \right) P^* - 125.$$

We have

$$0 = 0.05 \left(1 - \frac{P^*}{10,000} \right) P^* - 125.$$

On the right-hand side of the equation we have a quadratic:

$$0 = 0.05\left(1 - \frac{P^*}{10,000}\right)P^* - 125$$

$$0 = (0.05 - 0.000005P^*)P^* - 125$$

$$0 = -0.000005(P^*)^2 + 0.05P^* - 125.$$

Dividing both sides of the equation by the coefficient on the squared term yields the equation

$$0 = (P^*)^2 - 10,000P^* + 25,000,000.$$

We use the quadratic formula to solve:

$$P^* = \frac{-b \pm \sqrt{b^2 - 4ac}}{2a}$$

$$= \frac{10,000 \pm \sqrt{(-10,000)^2 - 4 \cdot 1 \cdot 25,000,000}}{2}$$

$$= \frac{10,000 \pm 0}{2}$$

$$= 5,000.$$

Because the discriminant turns out to be zero, we end up with only one solution. Our sole equilibrium value is $P^* = 5,000$.

b. Use Excel to confirm that the values found in a. are in fact equilibrium values.

To confirm that $P* = 5,000$ is an equilibrium value in Excel we need to enter the DDS and verify that if the DDS starts at 5,000, it stays at 5,000. Figure 1.28 shows the formula for the DDS along with the verification we need.

	A	B
1	Exercise 1.5.5	
2		
3	t	P(t)
4	0	5,000.00
5	1	=B4+0.05*(1-B4/10000)*B4-125
6	2	5,000.00
7	3	5,000.00
8	4	5,000.00
9	5	5,000.00
10	6	5,000.00
11	7	5,000.00
12	8	5,000.00

FIGURE 1.28 Excel confirmation of equilibrium value for Exercise 1.5.5.

c. Determine the stability of any equilibrium values found in a. by producing an appropriate Excel graph.

We graph the DDS for different starting points near the equilibrium value all on the same axes. Figure 1.29 shows the result. The behavior displayed in the figure is very different than what we are used to seeing. It appears that populations that start above 5,000 decrease back toward 5,000, but populations that start below 5,000 continue to decrease away from 5,000. This is an example of a **semi-stable** equilibrium value, which are discussed in Chapter 5. We also note that the horizontal line at 5,000 provides graphical confirmation that it is in fact an equilibrium value.

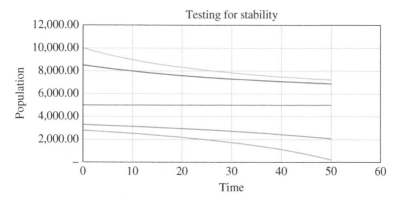

FIGURE 1.29   Excel graph testing for stability in Exercise 1.5.5.

7 Show that for *any* exponential model where there is no harvesting or stocking the only equilibrium value is 0.

The DDS for the general exponential model is given by $P(t) = P(t-1) + rP(t-1)$. We find all P^* such that $P^* = P^* + rP^*$. Then $0 = rP^*$, and we see that when $r \neq 0$, the only solution is $P^* = 0$. We can safely exclude the case where $r = 0$ since in that case we do not have a proper exponential model.

9 The U.S. Census in 2000 (see www.census.gov) estimated the population of the United States to be 281.4 million. Without immigration, the population would grow by approximately 0.6% each year. Data available at www.census.gov indicates that approximately 1,000,000 immigrants enter the U.S. each year.

a. Suppose that instead of growing by 0.6% per year the U.S. population was declining by 0.6% each year. Give the DDS for this situation.

The DDS is given by $P(t) = P(t-1) - 0.006P(t-1) + 1,000,000$.

b. At what value would the U.S. population stabilize in the long run?

Since this is an affine model the population would stabilize at the stable equilibrium value given by

$$P* = -\frac{a}{r} = -\frac{1{,}000{,}000}{-0.006} \approx 166{,}666{,}667.$$

The population would stabilize at about 167 million people.

c. Produce a graph that indicates the U.S. population would stabilize at this value no matter where it started.

This amounts to producing a graph that shows the equilibrium value is stable. Such a graph is given in Figure 1.30.

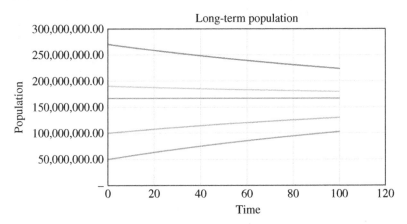

FIGURE 1.30 Excel graph testing for stability in Exercise 1.5.9.

d. If the U.S. government wanted to stabilize the population at 400,000,000, how many people should it allow to immigrate each year?

Here we have to arrange for the equilibrium value to be equal to 400,000,000. Assuming the same growth rate of −0.6%, we solve for a:

$$P^* = -\frac{a}{r}$$

$$400{,}000{,}000 = -\frac{a}{-0.006}$$

$$400{,}000{,}000 = \frac{a}{0.006}$$

$$2{,}400{,}000 = a.$$

We have found that if the U.S. population were declining by 0.6% each year and the U.S. government wanted a long-term population of 400,000,000, then it should allow approximately 2,400,000 people to immigrate to the U.S. each year.

2

PERSONAL FINANCE

2.1 COMPOUND INTEREST AND SAVINGS

1 Consider a savings account that earns 4% interest compounded monthly. You initially deposit $1,000.00 into the account and make no further deposits or withdrawals.

    a. Draw a flow diagram for the monthly balance in the account.

    Because interest is compounded monthly, each month the balance will increase by $\frac{4}{12}\%$. The flow diagram is given in Figure 2.1.

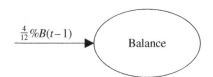

FIGURE 2.1   Flow diagram for Exercise 2.1.1.

    b. Find the DDS.

    The inward-pointing arrow represents an increase in the balance from the previous month, so the DDS is given by

Solutions Manual to Accompany Models for Life: An Introduction to Discrete Mathematical Modeling with Microsoft® Office Excel®, First Edition. Jeffrey T. Barton.
© 2016 John Wiley & Sons, Inc. Published 2016 by John Wiley & Sons, Inc.
Companion website: www.wiley.com/go/barton/solutionsmanual_modelsforlife

$$B(t) = B(t-1) + \frac{4}{12}\%B(t-1),$$

where time is in months.

c. Determine the account balance 2 years from the initial deposit.

Here we can use Excel or the explicit formula. To use the explicit formula we note that the initial balance is $B(0) = 1,000$, the monthly interest rate is $\frac{4}{12}\%$, and the time in months is $t = 24$. Plugging everything into the explicit formula gives

$$B(t) = \left(1 + \frac{r}{12}\right)^t B(0)$$

$$B(24) = \left(1 + \frac{0.04}{12}\right)^{24} 1,000$$

$$B(24) = 1,083.14.$$

The balance after 2 years, assuming no additional deposits or withdrawals, will be $1,083.14.

d. Determine how long it will take the deposit to double.

We set the balance equal to $2,000 and solve for the required time. First we have

$$2,000 = \left(1 + \frac{0.04}{12}\right)^t 1,000$$

$$2 = \left(1 + \frac{0.04}{12}\right)^t.$$

From here we can use a calculator and trial-and-error to find t, or we can take the natural logarithm of both sides:

$$\ln(2) = \ln\left[\left(1 + \frac{0.04}{12}\right)^t\right]$$

$$\ln(2) = t \cdot \ln\left(1 + \frac{0.04}{12}\right)$$

$$0.6931 \approx t \cdot 0.003328$$

$$208.3 \approx t.$$

It would take about 208 months, or about 17.4 years for the balance in the account to double.

3 Consider a savings account that earns 5% interest compounded monthly. Suppose you will withdraw $500 per month from the account.

a. Find the DDS for the balance.

Interest serves to increase the account balance while withdrawals decrease it. The DDS is given by $B(t) = B(t-1) + \frac{5}{12}\%B(t-1) - 500$, where the time units are months.

b. Find the equilibrium value for the balance.

Here we need to find B^* such that $B^* = B^* + \frac{5}{12}\%B^* - 500$. We have

$$0 = \frac{5}{12}\%B^* - 500$$
$$500 = 0.0041667B^*$$
$$120,000 = B^*.$$

The equilibrium balance for the account is $120,000.

c. Determine the stability of the equilibrium value.

Since this situation is equivalent to an exponentially increasing population with harvesting, we know that the equilibrium value is unstable. To confirm this we create a graph in Excel that shows the account balance over time for several different initial balances near $120,000. Figure 2.2 shows the result confirming that the equilibrium is unstable since all balances that start off of $120,000 are moving further away from it.

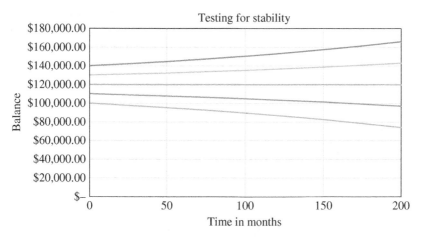

FIGURE 2.2 Excel graph testing for stability in Exercise 2.1.3.

d. Interpret the meaning of the equilibrium value in the context of a savings account balance.

The equilibrium value of $120,000 is the minimum balance we would need in the account in order to withdraw $500 per month indefinitely without running out of money.

5 The "Rule of 70" is a useful financial rule of thumb that estimates the doubling time for an investment. The Rule of 70 says that if r is the interest rate (given as a percent) for a savings account (or rate of return for an investment), then the time it takes for an initial deposit to double, assuming no further deposits or withdrawals, is approximately $\frac{70}{r}$ in years.

a. Compare what the rule estimates for doubling time with the actual doubling time for $r = 10\%$.

Rule of 70 estimate: $\frac{70}{10} = 7$ years.
Actual: Use natural logarithms to solve for the doubling time. (We could also use trial and error with a calculator.) Here we assume that the interest is compounded annually and that our time units are years.

$$B(t) = (1+r)^t B(0)$$
$$2B(0) = (1+0.10)^t B(0)$$
$$2 = (1.10)^t$$
$$\ln 2 = \ln\left[(1.10)^t\right]$$
$$0.6931 \approx t \cdot \ln(1.10)$$
$$0.6931 \approx t \cdot 0.0953$$
$$7.27 \approx t.$$

The actual doubling time is about 7.3 years, which is very close to the estimate given by the Rule of 70. Note that the result is independent of the initial deposit since the initial deposit ends up canceling. We could have used any value for $B(0)$ and we would get the same doubling time.

b. Compare what the rule estimates for doubling time with the actual doubling time for $r = 7\%$.

Rule of 70 estimate: $\frac{70}{7} = 10$ years.
Actual: Use natural logarithms to solve for the doubling time. (We could also use trial and error with a calculator.) Here we assume that the interest is compounded annually and that our time units are years.

$$B(t) = (1+r)^t B(0)$$
$$2B(0) = (1+0.07)^t B(0)$$
$$2 = (1.07)^t$$
$$\ln 2 = \ln\left[(1.07)^t\right]$$
$$0.6931 \approx t \cdot \ln(1.07)$$
$$0.6931 \approx t \cdot 0.0677$$
$$10.24 \approx t.$$

The actual doubling time is about 10.2 years, which is very close to the estimate given by the Rule of 70.

c. Compare what the rule estimates for doubling time with the actual doubling time for $r = 5\%$.

Rule of 70 estimate: $\frac{70}{5} = 14$ years.
Actual: Use natural logarithms to solve for the doubling time. (We could also use trial and error with a calculator.) Here we assume that the interest is compounded annually and that our time units are years.

$$B(t) = (1+r)^t B(0)$$
$$2B(0) = (1+0.05)^t B(0)$$
$$2 = (1.05)^t$$
$$\ln 2 = \ln\left[(1.05)^t\right]$$
$$0.6931 \approx t \cdot \ln(1.05)$$
$$0.6931 \approx t \cdot 0.0488$$
$$14.21 \approx t.$$

The actual doubling time is about 14.2 years, which is very close to the estimate given by the Rule of 70.

d. Compare what the rule estimates for doubling time with the actual doubling time for $r = 3\%$.

Rule of 70 estimate: $\frac{70}{3} \approx 23.33$ years.
Actual: Use natural logarithms to solve for the doubling time. (We could also use trial and error with a calculator.) Here we assume that the interest is compounded annually and that our time units are years.

$$B(t) = (1+r)^t B(0)$$
$$2B(0) = (1+0.03)^t B(0)$$
$$2 = (1.03)^t$$
$$\ln 2 = \ln\left[(1.03)^t\right]$$
$$0.6931 \approx t \cdot \ln(1.03)$$
$$0.6931 \approx t \cdot 0.0296$$
$$23.45 \approx t.$$

The actual doubling time is about 23.5 years, which is very close to the estimate given by the Rule of 70.

Overall the Rule of 70 seems to give a very good estimate for doubling times for a fairly wide range of reasonable interest rates.

7 Consider a savings account that earns 3% interest compounded monthly. Suppose you initially deposit $200 and subsequently deposit $50.00 per month.

a. Use Excel to find the balance in the account after 4 years.

We use our savings account spreadsheet where time is in months, and we need to drag the formula down to month 48 to find the balance in 4 years. The Excel formula and results are shown in Figure 2.3. Note that most of the rows are hidden. In 4 years the balance will be $2,772.03.

	A	B	C
1	Exercise 2.1.7		
2			
3	Interest rate, r =		3.0%
4	Monthly deposit, a =	$	50.00
5			
6	t	B(t)	
7	0 $	200.00	
8	1	=B7+($C$3/12)*B7+$C$4	
9	2 $	301.13	
54	47 $	2,715.24	
55	48 $	2,772.03	

FIGURE 2.3   Excel set-up and results for Exercise 2.1.7.

b. Use the appropriate explicit formula to find the same balance.

To find the same balance using the explicit formula we plug all known parameters into the formula:

$$B(t) = \left(1 + \frac{r}{12}\right)^t B(0) + a\frac{\left(1 + \frac{r}{12}\right)^t - 1}{r/12}$$

$$B(48) = \left(1 + \frac{0.03}{12}\right)^{48} 200 + 50\frac{\left(1 + \frac{0.03}{12}\right)^{48} - 1}{0.03/12}$$

$$B(48) \approx 225.47 + 50 \cdot 50.9312$$

$$B(48) \approx 2,772.03.$$

We get the same result as with Excel.

c. Over the entire 4 years, how much did you actually deposit into the account?

We made an initial deposit of $200, plus 48 deposits of $50.00 each. Thus the total deposited was $200 + 48 \times $50 = $2,600.00$.

d. By comparing your answer in c. to the account balance after 4 years, determine how much total interest you earned over the 4 years.

Since the account balance after 4 years is $2,772.03 but we only deposited $2,600.00, we must have earned $2,772.03 − $2,600.00 = $172.03 in interest.

9 *Extension:* Suppose you have found the world's greatest savings account, and it pays 100% interest. You deposit $1.00 into the account initially and make no further deposits or withdrawals.

a. Find the balance after 1 year if interest is paid annually.

$$B(1) = (1+1)^1 1.00$$
$$= 2.$$

The balance after 1 year will be $2.00.

b. Find the balance after 1 year if interest is compounded monthly.

With monthly compounding we use months as the time unit.

$$B(12) = \left(1 + \frac{1}{12}\right)^{12} 1.00$$
$$\approx 2.613.$$

After 1 year the account balance will be about $2.61.

c. Find the balance after 1 year if interest is compounded daily.

With daily compounding we use days as the time unit.

$$B(365) = \left(1 + \frac{1}{365}\right)^{365} 1.00$$
$$\approx 2.7146.$$

After 1 year the account balance will be about $2.71.

d. Find the balance after 1 year if interest is compounded hourly.

With hourly compounding we use hours as the time unit, and we need to know that there are $24 \times 365 = 8,760$ h in a year.

$$B(8,760) = \left(1 + \frac{1}{8,760}\right)^{8,760} 1.00$$
$$\approx 2.7181.$$

After 1 year the account balance will be about $2.72.

e. Find the balance after 1 year if interest is compounded every second.

With compounding every second we use seconds as the time unit, and we need to know that there are $60 \times 60 \times 24 \times 365 = 31,536,000$ s in a year.

$$B(31,536,000) = \left(1 + \frac{1}{31,536,000}\right)^{31,536,000} 1.00$$
$$\approx 2.71828.$$

After 1 year the account balance will be about $2.72.

f. Does there seem to be a limit on how high your balance can grow due to more frequent compounding? Do you recognize this limit?

While it is true that more frequent compounding is better for the savings account, there does seem to be a limit to how much more frequent compounding can help. In this example the limit is about an additional \$.72 over annual compounding. Note that we see almost all of the gains that more frequent compounding can provide once we are at daily compounding.

If we look at the last account balance when interest was compounded every second, we have the number 2.71828. This number is the first 6 digits of the irrational number e.

11 The \$1,000,000.00 goal we used for our retirement savings plan is based on someone who needs to generate \$40,000 a year to live on in retirement. Some people may need less income and others more. Take a few moments to decide how much income you would like to have to live on during your retirement years.

a. A reasonable rule of thumb is that you need to accumulate roughly 25 times your desired income in order to generate that income without dipping into your savings. Based on your income estimate and this rule of thumb, how much will you need to accumulate by the time you retire?

Suppose we want an income of \$55,000 per year in retirement. Then using the rule of thumb above, we would need to save $25 \times \$55,000 = \$1,375,000$ by the time we retire.

b. At what age would you like to retire?

Suppose we want to retire by age 60.

c. At what age do you foresee being able to start saving for retirement?

Assume we can start saving when we are 27 years old.

d. Assuming that your retirement account will earn the equivalent of 9% interest compounded monthly, determine how much you will need to save each month during your career.

Starting at age 27 and retiring at age 60 means we will be saving for 33 years, or 396 months. We could use Goal Seek to find the required monthly deposit, but proceeding as in Exercise 2.1.10 we note that with no initial deposit we can find a as follows:

$$a = \frac{B(t)}{\left[\dfrac{\left(1 + \frac{r}{12}\right)^t - 1}{r/12}\right]}$$

$$= \frac{1,375,000}{\left[\dfrac{\left(1 + \frac{0.09}{12}\right)^{396} - 1}{0.09/12}\right]}$$

$$\approx 564.23.$$

We would have to save \$564.23 per month.

13 *Extension:* As we advance in our careers, our income typically increases through periodic raises, and as our income rises, so should our retirement contributions. This is one good reason to base your retirement contribution on a percentage of your income (as in Exercise 2.1.12) instead of on a fixed amount over your entire career. Doing so allows you to increase your contribution over time without missing the extra money.

a. Set up an Excel spreadsheet for a retirement plan where the monthly contribution is a fixed percentage of your monthly salary. The following should be parameters stored in their own cells: monthly salary, monthly contribution %, and assumed interest rate.

The spreadsheet in Figure 2.4 shows the set-up with stand-in values chosen for the parameters. Note that the monthly contribution is the product of the monthly salary and the monthly contribution percentage.

	A	B	C
1	Exercise 2.1.13		
2			
3	Monthly salary =	$	3,000.00
4	Monthly contribution % =		15.0%
5	Assumed interest rate =		9.0%
6			
7	t	B(t)	
8	0	$ -	
9	1	=B8+($C$5/12)*B8+$C$3*$C$4	
10	2	$ 903.38	
11	3	$ 1,360.15	
12	4	$ 1,820.35	

FIGURE 2.4   Excel set-up for Exercise 2.1.13.

b. Assume you start saving 15% of your income each month at age 25 with a salary of $36,000 a year. If your account earns the equivalent of 9% interest compounded monthly, use your spreadsheet to compute the amount of money you will have saved by age 65.

With an annual salary of $36,000, the monthly salary is $3,000. Since we start saving at age 25 and continue to age 65, we save for a total of 40 years, or 480 months. We drag the formula down to month 480 and note the balance. Figure 2.5 shows the result with most of the rows hidden. We end up with about $2.1 million at age 65.

c. How much will you end up with if the situation is the same as in part b. except that now you get an annual 4% raise in salary?

There are several ways to approach this problem. First we note that a 4% raise in salary means a 4% increase in our monthly contribution. The tricky part is

	A	B	C
1	Exercise 2.1.13		
2			
3	Monthly salary =	$	3,000.00
4	Monthly contribution % =		15.0%
5	Assumed interest rate =		9.0%
6			
7	t	B(t)	
8	0	$	-
9	1	$	450.00
10	2	$	903.38
487	479	$2,090,465.63	
488	480	$2,106,594.12	

FIGURE 2.5   Excel output for Exercise 2.1.13b.

that the 4% increase only happens annually, i.e. every 12 months. One way to account for these annual increases is to manually edit the Excel formula every year to reflect the raise. This approach is somewhat tedious since we would have to modify the formula 39 different times.

Another, more efficient approach is to have separate columns for each year where we keep track of the monthly contribution for that year and therefore make it easier to incorporate the raises. Figure 2.6 shows the formula for the annual 4% increase in our monthly contribution, and that formula has been dragged *over* to year 40. In this way the monthly contribution that is in effect for each year is easily accessible.

Our time column now represents the month within each year. Thus the row for $t = 1$ represents the 1st month in year 1, the 1st month in year 2, etc. as we

	A	B	C	D	E
1	Exercise 2.1.13				
2					
3	Monthly salary =		$ 3,000.00		
4	Monthly contribution % =		15.0%		
5	Assumed interest rate =		9.0%		
6					
7	Year:		1	2	3	4
8	Month. Cont.: $	450.00	=B8+0.04*B8	$ 486.72	$ 506.19
9	t	B(t)	B(t)	B(t)	B(t)
10	0	$ -			
11	1	$ 450.00	6,138.63	12,568.02	19,815.96

FIGURE 2.6   Excel set-up for Exercise 2.1.13c.

move across the spreadsheet. The formula for the balance for the 1st month of each year is shown in Figure 2.7. That formula has been dragged over to year 40. Note that the formula takes the balance from the final month of the previous year, adds the interest, and then adds the new monthly contribution for that year.

	A	B	C
1	Exercise 2.1.13		
2			
3	Monthly salary =	$	3,000.00
4	Monthly contribution % =		15.0%
5	Assumed interest rate =		9.0%
6			
7	Year:	1	2
8	Month. Cont.: $	450.00 $	468.00
9	t	B(t)	B(t)
10	0 $	-	
11	1 $	450.00	=B22+($C$5/12)*B22+$C$8

FIGURE 2.7 Excel set-up for Exercise 2.1.13c continued.

Next Figure 2.8 shows the formula we use for the remaining 11 months in year 1. We drag that formula across to year 40, then drag all formulas down to month 12 simultaneously. Note in the formula that for the monthly contribution, the row number has a "$" in front of it but the column letter does not. This allows the column letter to update as we drag the formula across so it uses the correct monthly contribution for each year. However, when we copy the formulas down the row does not update, thus

	A	B	C
1	Exercise 2.1.13		
2			
3	Monthly salary =	$	3,000.00
4	Monthly contribution % =		15.0%
5	Assumed interest rate =		9.0%
6			
7	Year:	1	2
8	Month. Cont.: $	450.00 $	468.00
9	t	B(t)	B(t)
10	0 $	-	
11	1 $	450.00	6,138.63
12	2 $	903.38	=C11+($C$5/12)*C11+C$8

FIGURE 2.8 Excel set-up for Exercise 2.1.13c continued.

ensuring that the formula continues to use the correct cell for the monthly contribution.

Finally we check the balance in month 12 of year 40 and note that we end up with a balance of $3,180,280.55 at age 65. Figure 2.9 shows the result with most months and years hidden.

	A	B	AN	AO
1	Exercise 2.1.13			
2				
3	Monthly salary =			
4	Monthly contribution % =			
5	Assumed interest rate =			
6				
7	Year:	1	39	40
8	Month. Cont.: $	450.00 $	1,997.47 $	2,077.36
9	t	B(t)	B(t)	B(t)
10	0 $	-		
11	1 $	450.00	2,636,691.21	2,907,485.05
12	2 $	903.38	2,658,463.86	2,931,368.55
22	12 $	5,628.41	2,885,376.72	3,180,280.55

FIGURE 2.9   Excel output for Exercise 2.1.13d.

15 *Extension*: Following the spirit of the derivation for the explicit formula for an affine model, find the explicit formula for the general model described in Exercise 2.1.14. Confirm your result in 2.1.14d by using this explicit formula.

The derivation for the explicit formula works exactly the same as the one given in Exercise 1.4.13. The only difference is that instead of "r" we have "$\frac{r}{12}$." The final formula is given by

$$B(t) = \left(1 + \frac{r}{12}\right)^t B(0) + a_0 \left[\frac{(1+s)^t - \left(1 + \frac{r}{12}\right)^t}{s - \frac{r}{12}}\right].$$

To confirm our result in Exercise 2.1.14 we plug in all relevant parameters. With $B(0) = 400$, $\frac{r}{12} = \frac{0.03}{12} = 0.0025$, $s = 0.01$, $a_0 = 100$, and $t = 24$, we have

$$B(24) = (1.0025)^{24} 400 + 100 \left[\frac{(1.01)^{24} - (1.0025)^{24}}{0.01 - 0.0025}\right]$$

$$B(24) \approx 424.70 + 100 \cdot 27.7303$$

$$B(24) \approx 3,197.73.$$

We get the same result (with a very minor difference due to rounding) that we did with Excel.

2.2 BORROWING FOR MAJOR PURCHASES

1 Choose a car that you would like to buy, and use the Internet to find its cost and a source for current APRs for car loans. Assume that you have to finance the entire cost of the car.

Your Car (year, make and model): _____.
Price (give source): _____.
Source for APRs:_____.

Use Microsoft Excel to fill in the Text Table 2.2. The table should show what your monthly payment would be for loans of each term, and it should also show the total amount you paid for the car.

TEXT TABLE 2.2 Investigating the Effect of Loan Term on the Monthly Payment and Total Amount Paid for the Car

Number of Years to Repay	APR	Monthly Payment Required	Total Amount Paid for Car
3			
3.5			
4			
4.5			
5			
5.5			
6			

We give an example calculation for a term of 4 years here. The rest of the calculations are the same except that the loan term changes. According to Edmunds.com, we can expect to pay $36,772 for a new 2015 Audi A4. According to Bankrate.com, the average current APR for a 4-year new car loan is 2.95%.

To find the required monthly payment we can use Goal Seek in Excel or the explicit formula. We use the explicit formula here. With initial loan balance $B(0) = 36,772$, $r = 0.0295$, and $t = 48$, we have

$$B(t) = \left(1 + \frac{r}{12}\right)^t B(0) + a\frac{\left(1 + \frac{r}{12}\right)^t - 1}{r/12}$$

$$B(48) = \left(1 + \frac{0.0295}{12}\right)^{48} 36,772 + a\frac{\left(1 + \frac{0.0295}{12}\right)^{48} - 1}{0.0295/12}.$$

Since we need to pay off the loan at the end of 48 months (4 years), our balance after 48 months should be zero. Thus we have

$$0 = \left(1 + \frac{0.0295}{12}\right)^{48} 36,772 + a\frac{\left(1 + \frac{0.0295}{12}\right)^{48} - 1}{0.0295/12}.$$

Now we solve for a:

$$0 = \left(1 + \frac{0.0295}{12}\right)^{48} 36,772 + a\frac{\left(1 + \frac{0.0295}{12}\right)^{48} - 1}{0.0295/12}$$

$$0 \approx 41,371.485 + a \cdot 50.8805$$

$$\frac{-41,371.485}{50.8805} \approx a$$

$$-813.11 \approx a.$$

Our monthly payment will be $813.11.

To find the total amount paid for the car, we multiply the monthly payment by the total number of payments we make, which in this case is 48. We get

$$48 \times \$813.11 = \$39,029.28.$$

Thus the total amount we pay for the $36,772 car is $39,029.28.

3 Based on your table from Exercise 2.2.1, fill in the blanks:

a. As the length of the loan increases, the monthly payment _____.

Decreases. Taking out a longer loan generally results in lower monthly payments.

b. As the length of the loan increases, the total amount paid _____.

Increases. While the monthly payments are typically lower with a longer loan, the borrower has to make more payments, and as a result the total amount paid is generally higher than for shorter loans.

5 By making a down payment of 25%, how much do you save on the total amount paid for the car for a 5-year loan?

We continue our example using a 4-year term. With no down payment we pay $39,029.28 for the car. With a 25% down payment we pay $38,464.84 for the car. By making the down payment we save $39,029.28 − $38,464.84 = $564.44 on the purchase of the car.

7 The APR's you see advertised on television and the Internet are not always available to everyone. Typically these rates are reserved for customers who have excellent credit, and having a bad credit score can increase the APR you end up receiving. Suppose you borrow $20,000.00 with a 5-year term to purchase a new car. Determine how much additional total interest you pay on the loan for each additional percentage point in the APR. Consider APR's between 3% and 10%.

To find total interest paid on a loan, recall that we need to 1) find the monthly payment, 2) calculate the total loan cost by multiplying by the term in months, and 3) subtract the amount borrowed from the total loan cost. We need to do this 8 times: once for each whole number APR between 3% and 10%.

We modify the basic Excel loan spreadsheet to automate the process of finding total interest paid. The formula set-up is shown in Figure 2.10. For each APR, we need to run Goal Seek to find the monthly payment that results in a $0.00 balance in month 60. Then we record the results for each APR between 3% and 10%.

	A	B	C
1	Exercise 2.2.7		
2			
3	APR, r =		0.03
4	Monthly payment, a=		350.56
5	Total loan cost =		=60*C4
6	Total interest paid =		=C5-B9
7			
8	t (months)	Balance Owed	
9	0	20000	
10	=A9+1	=B9+($C$3/12)*B9-$C$4	

FIGURE 2.10   Excel set-up for Exercise 2.2.7.

• For an APR of 3%: total interest paid is $1,562.43.

The monthly payment for a $20,000, 5-year loan at 3% is found by Goal Seek (or the explicit formula) to be $359.37. Thus the total cost of the loan is $60 \times \$359.37 = \$21,562.20$. This means that the total interest paid on the loan is $\$21,562.20 - \$20,000.00 = \$1,562.20$. Figure 2.11 shows the complete Excel result. The minor difference in this versus our result is due to rounding in the monthly payment.

	A	B	C
1	Exercise 2.2.7		
2			
3	APR, r =		3.00%
4	Monthly payment, a=	$ 359.37	
5	Total loan cost =		$21,562.43
6	Total interest paid =		$ 1,562.43
7			
8	t (months)	Balance Owed	
9	0	$ 20,000.00	
10	1	$ 19,690.63	
68	59	$ 358.48	
69	60	$ (0.00)	

FIGURE 2.11   Excel output for Exercise 2.2.7.

• For an APR of 4%: total interest paid is $2,099.83.
• For an APR of 5%: total interest paid is $2,645.48.

- For an APR of 6%: total interest paid is $3,199.36.
- For an APR of 7%: total interest paid is $3,761.44.
- For an APR of 8%: total interest paid is $4,331.67.
- For an APR of 9%: total interest paid is $4,910.03.
- For an APR of 10%: total interest paid is $5,496.45.

Though the difference depends on the starting APR, every additional percentage point in the APR leads to an additional $537.40 to $586.42 in interest charges for a 5-year $20,000 loan.

9   A standard down payment for a home mortgage is 20% of the price of the house. Repeat Exercises 2.2.8d–f assuming you make a down payment of 20%. How much interest do you save by making the down payment?

We re-run our Goal Seek command after entering 20% for our down payment percentage. The new PIP is $1,069.43, which gives a total loan cost of

$$360 \times \$1,069.43 = \$384,994.80.$$

Thus the total cost of buying the house is

$$\$384,994.8 + \$55,000.00 = \$439,994.80.$$

Figure 2.12 shows the Excel details and part of the amortization schedule. The total interest paid with no down payment is

$$\$481,240.80 - \$275,000.00 = \$206,240.80.$$

	A	B	C	D	E
1	Exercise 2.2.10				
2					
3	Loan with no Points				
4					
5	APR, $r =$		4.15%		
6	Purchase price =		$275,000.00		
7	Down payment % =		20.00%		
8	Down payment =		$55,000.00		
9	Monthly PIP, $a=$		$1,069.43		
10					
					Total
			Interest	Principal	Interest to
11	t (months)	Balance Owed	Payment	Payment	Date
12	0	$220,000.00			
13	1	$219,691.41	$760.83	$308.59	$760.83
371	359	$1,065.74	$7.36	$1,062.07	$164,990.15
372	360	$0.00	$3.69	$1,065.74	$164,993.83

FIGURE 2.12   Excel output for Exercise 2.2.9.

The total interest paid with the down payment is

$$\$384,994.80 - \$220,000.00 = \$164,994.80.$$

Thus by making a down payment of $55,000.00 we end up saving $41,246 in interest charges over the life of the loan.

11 Based purely on total cost over the full 30 years, which is preferable: paying the 2 points or not? (Assume no down payment in either case.)

The total house cost without points and an APR of 4.15% was $481,240.80. The total house cost with paying 2 points for an APR of 3.65% was $458,383.60. Paying the points saves us $22,857.20 over the life of the loan and is therefore preferable.

13 How much total do you pay for the loan with no down payment and 2 points if you completely pay the loan off after 6 months?

When we pay points, we have to add the cost of those points to the interest paid in order to get the total loan cost. Figure 2.13 shows the Excel details. We end up paying $10,499.44 in points and interest if we pay off the loan in 6 months.

F	G	H	I
Exercise 2.2.13			
Loan with Points			
Points Paid =		$5,500.00	
APR, r =		3.65%	
Purchase price =		$275,000.00	
Down payment % =		0.00%	
Down payment =		$0.00	
Monthly PIP, a=		$1,258.01	
			Total Interest to
	Interest	Principal	Date Plus
Balance Owed	Payment	Payment	Points
$275,000.00			
$274,578.44	$836.46	$421.56	$6,336.46
$274,155.61	$835.18	$422.84	$7,171.63
$273,731.48	$833.89	$424.12	$8,005.52
$273,306.07	$832.60	$425.41	$8,838.12
$272,879.36	$831.31	$426.71	$9,669.43
$272,451.36	$830.01	$428.01	$10,499.44

FIGURE 2.13   Excel output for Exercise 2.2.13.

15 At what point would paying each loan off completely result in equal costs between paying no points and paying 2 points?

Here we have to compare amortization schedules for both loans side-by-side. We look for the month in which the total loan costs to date are the same for both loans. Figure 2.14 shows the result of our comparison for the 30-year, $275,000 loan at 4.15% with no points and at 3.65% with 2 points. As Figure 2.14 indicates, there is no month where the total loan costs are identical; however, we note that between months 48 and 49, the loan with points goes from costing more to costing less to date than the loan with no points. When paying off the loan early before month 49, the loan with no points is the better choice, while after month 49 the loan with points is the better choice. The closest we can get to equal costs is month 49.

					Loan with Points			
3	Loan with no Points				Loan with Points			
4					Points Paid =		$5,500.00	
5	APR, r =		4.15%		APR, r =		3.65%	
6	Purchase price =		$275,000.00		Purchase price =		$275,000.00	
7	Down payment % =		0.00%		Down payment % =		0.00%	
8	Down payment =		$0.00		Down payment =		$0.00	
9	Monthly PIP, a=		$1,336.78		Monthly PIP, a=		$1,258.01	
10								
				Total				Total
			Interest	Principal	Interest to		Interest	Principal	Date Plus
11	t (months)	Balance Owed	Payment	Payment	Date	Balance Owed	Payment	Payment	Points
60	48	$254,896.58	$883.09	$453.70	$44,062.22	$253,249.15	$771.78	$486.24	$44,133.82
61	49	$254,441.32	$881.52	$455.27	$44,943.74	$252,761.44	$770.30	$487.71	$44,904.11
62	50	$253,984.48	$879.94	$456.84	$45,823.68	$252,272.24	$768.82	$489.20	$45,672.93
63	51	$253,526.05	$878.36	$458.42	$46,702.05	$251,781.56	$767.33	$490.69	$46,440.26

FIGURE 2.14   Excel output for Exercise 2.2.15.

2.3   CREDIT CARDS

For the following exercises, use the credit card with terms given in the abbreviated Schumer box in Text Table 2.4.

TEXT TABLE 2.4   Schumer Box for Section 2.3 Exercises

Interest Rates and Interest Charges	
APR for purchases	18.9%
APR for balance transfers	21.9%
APR for cash advances	24.9%
Penalty APR	28.9%
Grace period	25 days from end of billing cycle
Minimum interest charge	$1.00

Fees	
Annual Fee	$45.00
Balance transfer fee	$10.00 or 4%, whichever is greater
Cash advance fee	$10.00 or 5%, whichever is greater
Foreign purchase transaction fee	None
Late Payment	$35.00
Returned Payment	$35.00

1 Suppose you took out a $25,000 car loan for a term of 60 months using this credit card's cash advance APR as the APR for the car loan. How much in total would that car end up costing you?

The cash advance APR for the credit card is 24.9%. We take 24.9% as the APR for our car loan and use it to find the monthly payment that would pay off the car loan in 60 months. Using the explicit formula for car loans allows us to solve for the monthly payment:

$$B(t) = \left(1 + \frac{r}{12}\right)^t B(0) + a \frac{\left(1 + \frac{r}{12}\right)^t - 1}{r / 12}$$

$$0 = \left(1 + \frac{0.249}{12}\right)^{60} 25,000 + a \frac{\left(1 + \frac{0.249}{12}\right)^{60} - 1}{0.249 / 12}$$

$$0 \approx 85,724.177 + a \cdot 117.059$$

$$\frac{-85,724.177}{117.059} \approx a$$

$$-732.32 \approx a.$$

The total amount paid for the car is the monthly payment times the number of payments, in this case 60. The result is $43,939.20, about 1.75 times the price of the car.

3 Suppose you transfer a balance of $1,200 from another credit card to this one. Assuming you make no other payments or transfers, how much would you owe at the end of 1 month?

If we transfer a balance we first have to account for any balance transfer fees. For this card that is the greater of $10.00 and 4% of $1,200. Since 4% of $1,200 is $48.00, the fee here will be $48.00. Thus as soon as we transfer the $1,200 balance, we actually owe $1,248.00 on the card. Using $t = 30$ days for 1 month, we use daily compounding with the balance transfer APR of 21.9% in the explicit formula:

$$B(t) = \left(1 + \frac{r}{365}\right)^t B(0)$$

$$B(30) = \left(1 + \frac{0.219}{365}\right)^{30} 1,248.00$$

$$B(30) \approx 1,270.66.$$

At the end of 1 month we owe the credit card company $1,270.66. Had we instead used 31 days for 1 month, the balance would be slightly higher at $1,271.42.

2.4   THE TIME VALUE OF MONEY: PRESENT VALUE

For all problems assume access to an account that has an interest rate of $r = 5.5\%$ compounded annually.

1   Find the present value for a future payment of $5,000.00 to be made 6 years from today. Explain what this present value represents in a complete sentence.

The present value of the future payment of $5,000.00 is the amount we would have to deposit into our account earning 5.5% in order to have $5,000.00 in the account in 6 years. Thus we need to solve for $B(0)$ so that $B(6) = 5,000$. We have

$$B(t) = (1+r)^t B(0)$$

$$5,000 = (1+0.055)^6 B(0)$$

$$\frac{5,000}{1.055^6} = B(0)$$

$$3,626.23 \approx B(0).$$

The present value of the $5,000.00 payment in 6 years is $3,626.23. This is the amount we would need to put into our account today in order to have $5,000.00 in 6 years.

3   Suppose you need $50,000 per year to live comfortably in retirement. If you expect your retirement to last for 30 years, determine the nest egg you will need on the day you retire. Explain what your result means in the context of the problem.

This problem requires us to find the amount of money we need in our account at the beginning of our retirement so that we can exactly generate 30 annual payments of $50,000. Equivalently, we need the amount required in our account when we retire so that we can make 30 annual withdrawals of $50,000.00 each and end up with a zero balance. We use the explicit formula to solve for $B(0)$, where $B(30) = 0$, $r = 0.055$, $t = 30$, and $a = -50,000$. We have

$$B(t) = (1+r)^t B(0) + a\frac{(1+r)^t - 1}{r}$$

$$0 = (1.055)^{30} B(0) - 50,000\frac{(1.055)^{30} - 1}{0.055}$$

$$0 \approx 4.984 B(0) - 50,000 \cdot 72.435$$

$$\frac{3,621,773.90}{4.984} \approx B(0)$$

$$726,680.16 \approx B(0).$$

In order to fund our 30-year retirement we need to have saved $726,680.16 by the time we retire.

5 *Extension*: A **perpetuity** is a series of future payments that continues forever. Use the concept of an equilibrium value to find the present value of a perpetuity that pays $50,000 every year starting 1 year from today.

We need to find the amount required to be put into an account today in order to generate payments of $50,000 per year indefinitely. In order for the account to never run out of money, we have to make sure that the account balance does not decrease with each payment. This means at a minimum we want the account to be at equilibrium.

We need to find B^* such that

$$B^* = B^* + 0.055B^* - 50,000$$
$$0 = 0.055B^* - 50,000$$
$$\frac{50,000}{0.055} = B^*$$
$$909,090.91 \approx B^*.$$

Thus in order to generate an infinite series of annual payments of $50,000, we need to deposit $909,090.91 into an account today. We say that the present value of the perpetuity is $909,090.91.

7 *Extension*: In recent years lotteries have changed the way they handle the annuity payout option. Lotteries now use increasing annuities where the annual payment increases by a fixed percentage every year. Powerball uses an increase of 4% every year while Mega Millions uses an increase of 5% every year. (So *if* the Powerball annuity payout started with a $1,000,000 payment, the second annual payment would be $1,040,000.) The annual payments still begin with the first payment today and then continue for the next 29 years. The sum of the annual payments must still equal the jackpot.

a. Determine the annual payouts for Powerball if the sum of all 30 annual payouts must equal the jackpot of $50 million.

First we modify our spreadsheet for the present value of irregular payments to automatically increase the annual payments by 4% per year. With stand-in values for the first annual payment and for the present value, we show the set-up and formula in Figure 2.15.

	A	B	C
1	Exercise 2.4.7		
2			
3	Assumed interest rate, *r* =		4.50%
4	Present value =		$35,000,000.00
5			
6	*t*	Balance	Future Payments
7	0	$34,000,000.00	$1,000,000.00
8	1	$34,490,000.00	=C7+0.04*C7
9	2	$34,960,450.00	$1,081,600.00
10	3	$35,408,806.25	$1,124,864.00
11	4	$35,832,343.97	$1,169,858.56

FIGURE 2.15 Excel set-up for Exercise 2.4.7.

Next we find the actual annual payments by using Goal Seek. Note that the first payment is the important one to find since the rest of the payments are all based on 4% increases from then on. Thus we use Goal Seek to find the first payment that makes the sum of all 30 equal the jackpot of $50,000,000. To find the sum of all 30 payments we make use of Excel's SUM command as shown in Figure 2.16, though we could also manually enter the sum. The command adds together all cells between C7 and C36.

	A	B	C
1	Exercise 2.4.7		
2			
3	Assumed interest rate, $r =$		4.50%
4	Present value =		$35,000,000.00
5			
6	t	Balance	Future Payments
7	0	$34,000,000.00	$1,000,000.00
8	1	$34,490,000.00	$1,040,000.00
35	28	$26,962,491.66	$2,998,703.32
36	29	$25,057,152.33	$3,118,651.45
37			=SUM(C7:C36)

FIGURE 2.16   Use of Excel's SUM command for Exercise 2.4.7.

Next the set-up of our Goal Seek command is shown in Figure 2.17. The resulting annual payments (with most rows hidden) are given in Figure 2.18.

	A	B	C	D	E
1	Exercise 2.4.7				
2					
3	Assumed interest rate, $r =$		4.50%		
4	Present value =		$35,000,000.00		
5					
6	t	Balance	Future Payments		
7	0	$34,000,000.00	$1,000,000.00		
8	1	$34,490,000.00	$1,040,000.00		
35	28	$26,962,491.66	$2,998,703.32		
36	29	$25,057,152.33	$3,118,651.45		
37			$56,084,937.75		

Goal Seek
Set cell: C37
To value: 50,000,000
By changing cell: $C$7
OK   Cancel

FIGURE 2.17   Excel Goal Seek set-up for Exercise 2.4.7a.

b. Determine the present value of the annuity option for Powerball.

Now that we know the annual payments, we can use Goal Seek to find the present value that would result in $B(29) = 0$. The set-up just before running Goal Seek is given in Figure 2.19. The result is a present value of $24,969,875.48.

	A	B	C
1	Exercise 2.4.7		
2			
3	Assumed interest rate, $r =$		4.50%
4	Present value =		$35,000,000.00
5			
6	t	Balance	Future Payments
7	0	$34,108,495.04	$891,504.96
8	1	$34,716,212.17	$927,165.15
35	28	$37,060,885.68	$2,673,358.87
36	29	$35,948,332.31	$2,780,293.23
37			$50,000,000.00

FIGURE 2.18   Excel Goal Seek result for Exercise 2.4.7a.

	A	B	C	D	E
1	Exercise 2.4.7				
2					
3	Assumed interest rate, $r =$		4.50%		
4	Present value =		$35,000,000.00		
5					
6	t	Balance	Future Payments		
7	0	$34,108,495.04	$891,504.96		
8	1	$34,716,212.17	$927,165.15		
35	28	$37,060,885.68	$2,673,358.87		
36	29	$35,948,332.31	$2,780,293.23		
37			$50,000,000.00		

Goal Seek

Set cell: B36
To value: 0
By changing cell: $C$4

OK   Cancel

FIGURE 2.19   Excel Goal Seek set-up for Exercise 2.4.7b.

c. Which annuity is better from a present value point of view: level payments or increasing payments? Note that the answer to this question suggests a different reason for why lotteries may have changed their payout method.

The present value of level annual payments was found in Exercise 2.4.6 to be $28,369,935.58, thus level payments would result in a higher present value than increasing annual payments. This is due to the fact that with level payments more of the money is received sooner.

9 Given equal jackpots, which game would you prefer to win and why – Powerball or Mega Millions?

With a jackpot of $50 million, the present value of the increasing annuity option for Powerball is about $750,000 greater than the present value of the increasing annuity option for Mega Millions. Given a choice, it would be better to win the Powerball lottery given equal jackpots.

2.5   CAR LEASES

1 Choose a vehicle's make and model, and use www.swapalease.com to find a lease
   takeover for that vehicle. Note that you should refine the search to only include
   takeovers that include the cost of the optional lease buyout.

   a. Find the present value of the lease takeover by finding the present value of all
   future payments including the optional lease buyout. You will need to supply
   your own assumed interest rate and note how you determined it.

   For example purposes, we select a 2013 BMW 3 Series that has 13 months
   remaining on its lease. The monthly payment is $436.56 and the optional lease
   end buyout is $26,160.00. We assume that we have access to a savings account
   that earns 3% interest compounded monthly.

   We need to find how much money we would need to put into our account
   today to cover the remaining lease payments and purchase the car at the end of
   the lease. This will be the present value of the lease takeover. We use the Excel
   spreadsheet we developed for finding the present value of a lease takeover. The
   set-up is given in Figure 2.20 with a stand-in value for the present value.

   Next we use Goal Seek to find the present value that results in a balance of
   $0.00 in month 13. The set-up just before running Goal Seek is shown in
   Figure 2.21. The result of a successful Goal Seek is the present value of the
   lease takeover, in this case $30,915.63.

	A	B	C
1	Exercise 2.5.1		
2			
3	Assumed interest rate, r =		3.00%
4	Present value =		$40,000.00
5			
6	t	Balance	Future Payments
7	0	$39,563.44	$436.56
8	1	$39,225.79	$436.56
9	2	$38,887.29	$436.56
10	3	$38,547.95	$436.56
11	4	$38,207.76	$436.56
12	5	$37,866.72	$436.56
13	6	$37,524.83	$436.56
14	7	$37,182.08	$436.56
15	8	$36,838.47	$436.56
16	9	$36,494.01	$436.56
17	10	$36,148.69	$436.56
18	11	$35,802.50	$436.56
19	12	$35,455.44	$436.56
20	13	$9,384.08	$26,160.00

FIGURE 2.20   Excel set-up for Exercise 2.5.1.

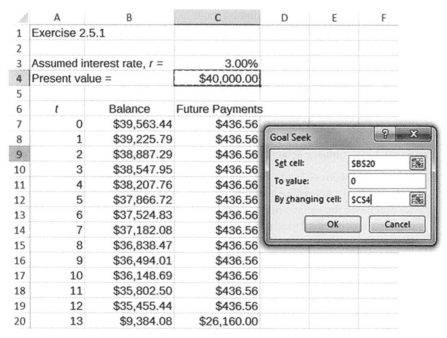

	A	B	C	D	E	F
1	Exercise 2.5.1					
2						
3	Assumed interest rate, r =		3.00%			
4	Present value =		$40,000.00			
5						
6	t	Balance	Future Payments			
7	0	$39,563.44	$436.56			
8	1	$39,225.79	$436.56			
9	2	$38,887.29	$436.56			
10	3	$38,547.95	$436.56			
11	4	$38,207.76	$436.56			
12	5	$37,866.72	$436.56			
13	6	$37,524.83	$436.56			
14	7	$37,182.08	$436.56			
15	8	$36,838.47	$436.56			
16	9	$36,494.01	$436.56			
17	10	$36,148.69	$436.56			
18	11	$35,802.50	$436.56			
19	12	$35,455.44	$436.56			
20	13	$9,384.08	$26,160.00			

Goal Seek

Set cell: $B$20

To value: 0

By changing cell: $C$4

OK   Cancel

FIGURE 2.21   Excel Goal Seek set-up for Exercise 2.5.1.

Customized True Market Value® Prices

	Trade-In	Private Party	Dealer Retail
National Base Price	$21,887	$23,707	$25,196
Optional Equipment	$2,283	$2,443	$2,859
Power Passenger Seat Lumbar Adjustment	$52	$56	$66
18 Inch Alloy Wheels	$428	$458	$535
Power Glass Sunroof	$474	$507	$594
Power Driver's Seat Lumbar Adjustment	$52	$56	$66
Anti-Theft Alarm System	$182	$195	$228
Adaptive Cruise Control	$1,095	$1,171	$1,370
Color Adjustment - White	$66	$71	$76
Regional Adjustment - for Zip Code 77024	$28	$31	$33
Mileage Adjustment - 19,130 miles	$916	$916	$916
Condition Adjustment - Clean	$0	$0	$0
Total	$25,180	$27,168	$29,080

Buying a Certified Used Vehicle	Dealer Retail
Certified Used Price	$31,186

FIGURE 2.22   Edmunds.com used car appraisal for Exercise 2.5.1.

b. Use a website such as www.edmunds.com or www.kbb.com to determine the amount you would have to pay for the same car today if you were to buy it outright. Be sure to include all options when determining the car's value.

Here we are appraising the exact car we are considering in order to determine how much it would cost us to purchase it outright rather than taking over the lease. We use www.edmunds.com to appraise the vehicle, and it is important that we are careful to include all options as stated in the Swapalease.com ad. Figure 2.22 shows the result of our appraisal.

c. Explain in a complete sentence or two whether the lease takeover is a good deal.

This lease takeover has a present value of $30,915.63 and appears to be a decent deal. According to Edmunds.com we could probably do better if we purchased the vehicle from a private seller or even from a dealer. However, a comparable certified used car would be slightly more expensive than the lease takeover option.

3

COMBAT MODELS

3.1 LANCHESTER COMBAT MODEL

1 Suppose Blue begins a battle with 200 units, so $B(0) = 200$, and Red begins the battle with 100 units, so $R(0) = 100$. Each Blue unit has a fighting effectiveness of $b = 0.10$, and each Red unit has a fighting effectiveness of $r = 0.15$.

    a. Determine by hand the numbers of Blue and Red forces remaining after two time steps.

According to the Lanchester combat model, the number of Blue forces remaining after one time step is given by

$$B(1) = B(0) - rR(0)$$
$$B(1) = 200 - 0.15 \cdot 100$$
$$B(1) = 185.$$

The number of Red forces remaining after one time step is given by

$$R(1) = R(0) - bB(0)$$
$$R(1) = 100 - 0.10 \cdot 200$$
$$R(1) = 80.$$

Solutions Manual to Accompany Models for Life: An Introduction to Discrete Mathematical Modeling with Microsoft® Office Excel®, First Edition. Jeffrey T. Barton.
© 2016 John Wiley & Sons, Inc. Published 2016 by John Wiley & Sons, Inc.
Companion website: www.wiley.com/go/barton/solutionsmanual_modelsforlife

Now that we have the force levels after one time step, we apply the model equations again to find the force levels after 2 time steps. For Blue we get

$$B(2) = B(1) - rR(1)$$
$$B(2) = 185 - 0.15 \cdot 80$$
$$B(2) = 173.$$

For Red we get

$$R(2) = R(1) - bB(1)$$
$$R(2) = 80 - 0.10 \cdot 185$$
$$R(2) = 61.5.$$

After 2 time steps there will be 173 Blue forces and 61.5 Red forces.

b. Determine the eventual victor and how many forces remain at the end of the battle.

Here we use the Lanchester Excel spreadsheet. We input the relevant parameters and copy the formulas down until one side has no forces remaining. The results are shown in Figure 3.1. Blue wins after 6 time steps and has 151.2 forces remaining.

	A	B	C
1	Exercise 3.1.1		
2			
3	Fighting Effectiveness:		
4	For Blue, b =		0.10
5	For Red, r =		0.15
6			
7	t	$B(t)$	$R(t)$
8	0	200	100
9	1	185.0	80.0
10	2	173.0	61.5
11	3	163.8	44.2
12	4	157.1	27.8
13	5	153.0	12.1
14	6	151.2	0.0

FIGURE 3.1   Excel output for Exercise 3.1.1.

3 For the parameters given in Exercise 3.1.1, determine the following:

a. The relative losses sustained by each side during the first time step.

The relative loss sustained by Blue during the first time step is given by $\frac{\Delta B}{B}$, where ΔB represents the number of units lost during the first time step and B is the initial number of Blue units. From Exercise 3.1.1 we know that Blue lost 15 units during the first time step, so

$$\frac{\Delta B}{B} = \frac{15}{200} = 0.075.$$

For Red the relative loss is given by

$$\frac{\Delta R}{R} = \frac{20}{100} = 0.20.$$

b. The fractional exchange ratio, FER.

The fractional exchange ratio is given by

$$FER = \frac{\Delta B/B}{\Delta R/R} = \frac{0.075}{0.20} = 0.375.$$

c. The fighting strengths of each side.

The fighting strength for Blue is given by

$$bB(0)^2 = 0.10 \cdot 200^2 = 4,000.$$

The fighting strength for Red is given by

$$rR(0)^2 = 0.15 \cdot 100^2 = 1,500.$$

d. The eventual victor based on the FER.

Since $FER = 0.375 < 1$, Theorem 3.1 tells us that Blue will eventually win.

e. The eventual victor based on the fighting strengths.

Since Blue's fighting strength is greater than Red's fighting strength, we know that Blue will eventually win.

f. How do your answers in d. and e. compare to the result found in Exercise 3.1.1b.?

Both agree with the numerical result found with Excel in Exercise 3.1.1b.

5   Suppose Red has 200 units and a fighting effectiveness of $r = 0.20$. If Blue has a fighting effectiveness of $b = 0.25$, how many forces would Blue need initially to completely put Red out of action in the first time step?

In order for Blue to put Red completely out of action in the first time step, we must have $R(0) - bB(0) \leq 0$. Thus we need

$$200 - 0.25 \cdot B(0) \leq 0$$
$$200 \leq 0.25 \cdot B(0)$$
$$800 \leq B(0).$$

If Blue starts the battle with at least 800 units, Red will be completely put out of action in the first time step.

7 Suppose that 500 Blue units engage 420 Red units where the Red units are more effective fighters with $b = 0.03$ and $r = 0.07$.

   a. Use fighting strength to predict the winner.

   Blue's fighting strength is $bB(0)^2 = 0.03 \cdot 500^2 = 7,500$.
   Red's fighting strength is $rR(0)^2 = 0.07 \cdot 420^2 = 12,348$.
   Since Red has the greater fighting strength, Red will win.

   b. Use Lanchester's Square Law to estimate the number of units remaining for the winner.

   From the discussion of Lanchester's Square Law we know that for all t,

   $$rR(t)^2 - bB(t)^2 \approx rR(0)^2 - bB(0)^2.$$

   Since we know Red will win, eventually we have $B(t) = 0$ so that

   $$0.07 \cdot R(t)^2 - 0 \approx 0.07 \cdot 420^2 - 0.03 \cdot 500^2$$
   $$0.07 \cdot R(t)^2 \approx 4,848$$
   $$R(t)^2 \approx 69,257.14$$
   $$R(t) \approx \sqrt{69,257.14} \approx 263.17.$$

   At the end of the battle we predict that Red will have about 263.17 units remaining.

   c. Compare your results in a. and b. to the Lanchester Excel model projections.

   Plugging the relevant parameters into Excel we get the results shown in Figure 3.2 (with most rows hidden). The Excel results confirm that Red wins

	A	B	C
1	Exercise 3.1.7		
2			
3	Fighting Effectiveness:		
4	For Blue, b =		0.03
5	For Red, r =		0.07
6			
7	t	$B(t)$	$R(t)$
8	0	500	420
9	1	470.6	405.0
30	22	13.7	257.3
31	23	0.0	256.9

FIGURE 3.2 Excel output for Exercise 3.1.7.

the battle, and we see that Red will have 256.9 units remaining. This compares favorably with our estimate of 263.17 above.

9 *Extension*: Suppose there is a three-way battle among Red, Blue, and Green forces. Develop a modification of the Lanchester model for this situation. You may want to introduce additional parameters.

Let b, r, and g be the fighting effectiveness for Blue, Red, and Green, respectively. If we imagine a battle in which Blue, Red, and Green all attack each other, then as a first pass our model might look as follows:

$$B(t) = B(t-1) - rR(t-1) - gG(t-1)$$

$$R(t) = R(t-1) - bB(t-1) - gG(t-1)$$

$$G(t) = G(t-1) - bB(t-1) - rR(t-1).$$

The idea here is that each force suffers losses at the hands of the other two.

A problem with this set-up is that it assumes that each side inflicts the same number of casualties on the other two based on the fighting effectiveness parameter. A more realistic approach would be to allow each side to concentrate more or less of its firepower on each of its enemies. In such a set-up we would need new parameters indicating the proportion of its firepower each side will devote to the other two.

As an example, Green may want to devote 90% of its offensive firepower to attacking Blue and only 10% to attacking Red. Blue may want to devote 60% of its firepower to attacking Red and 40% to attacking Green, and Red may want to devote 70% of its firepower to attacking Blue and 30% to Green. In this scenario our model would become

$$B(t) = B(t-1) - 0.70 \cdot rR(t-1) - 0.90 \cdot gG(t-1)$$

$$R(t) = R(t-1) - 0.60 \cdot bB(t-1) - 0.10 \cdot gG(t-1)$$

$$G(t) = G(t-1) - 0.40 \cdot bB(t-1) - 0.30 \cdot rR(t-1).$$

The advantage of the new model is that it includes a kind of strategic parameter that keeps the number of casualties inflicted by a side consistent with the basic Lanchester model but allows each side to decide how to distribute those casualties between its two enemies.

11 *Extension*: Suppose that like Admiral Nelson, Blue aims to employ a divide and conquer strategy against Red in an upcoming battle. Show that dividing the Red forces into two equal halves minimizes the total fighting strength of the two divided forces.

Suppose we divide Red into two forces: R_1 and R_2. Let x be the number of forces in R_1, where $0 \leq x \leq R(0)$. We do not require x to be a whole number. Then

$R_1(0) = x$, and $R_2(0) = R(0) - x$. The total fighting strength of the divided forces for Red is given by

$$rR_1(0)^2 + rR_2(0)^2 = rx^2 + r(R(0) - x)^2.$$

Our goal is to show that the right-hand side of the above equality attains its minimum value when Red is split into two equal forces, i.e. when $x = \dfrac{R(0)}{2}$.

We begin by expanding the right-hand side to get

$$rx^2 + r(R(0) - x)^2 = rx^2 + r\left(R(0)^2 - 2R(0)x + x^2\right).$$

Collecting like terms on the right and factoring gives

$$rx^2 + r\left(R(0)^2 - 2R(0)x + x^2\right) = r\left(2x^2 - 2R(0)x + R(0)^2\right)$$

$$= 2r\left(x^2 - R(0)x + \frac{R(0)^2}{2}\right).$$

Finally we complete the square inside the parentheses to get

$$2r\left(x^2 - R(0)x + \frac{R(0)^2}{2}\right) = 2r\left[\left(x - \frac{R(0)}{2}\right)^2 + \frac{R(0)^2}{4}\right].$$

Since $2r$ must be positive, the expression on the right attains its minimum value whenever $\left(x - \dfrac{R(0)}{2}\right)^2 = 0$, which can only occur when $x = \dfrac{R(0)}{2}$. Thus we have shown that if a force is to be divided into two smaller forces, the total fighting strength of the divided forces is minimized when the original force is divided into two equal forces.

13 *Extension*: Determine the limit for how much a divide and conquer strategy can diminish a force's total fighting strength. To do so consider an initial force of N units divided into N forces of 1 unit each. In other words, each unit must fight the enemy alone.

  If we have N individual forces, then their total fighting strength is the sum of their individual fighting strengths. Considering Blue as our example, each force has fighting strength equal to $b \cdot 1^2 = b$. Since we have N individual forces the total sum will be $N \times b = bN$. Though we do not offer a proof, we have found the minimum possible fighting strength for a force of size N as a result of a divide and conquer strategy.

3.2   PHASE PLANE GRAPHS

1   Suppose Blue begins a battle with 200 units, so $B(0) = 200$, and Red begins the battle with 100 units, so $R(0) = 100$. Each Blue unit has a fighting effectiveness of $b = 0.10$, and each Red unit has a fighting effectiveness of $r = 0.15$.

a. Determine the eventual victor by plotting the trajectory of the battle in the B,R-phase plane.

We use the Excel spreadsheet for graphing a battle trajectory for the Lanchester combat model, being careful to enter the correct parameters. The result is shown in Figure 3.3, and the graph indicates that Blue will eventually win the battle.

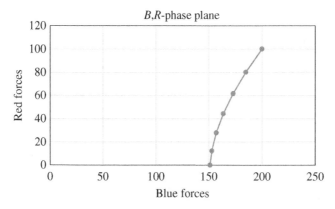

FIGURE 3.3   Excel phase plane trajectory for Exercise 3.2.1a.

b. Explain how the result in a. indicates the eventual victor.

The trajectory starts at the point $(\text{Blue}, \text{Red}) = (200,100)$ and ends on the Blue axis at approximately the point $(\text{Blue}, \text{Red}) = (150,0)$. Since Red ends up with 0 forces, Blue is the victor.

c. Produce a phase plane diagram showing the fate of several battles for different initial numbers of Blue and Red forces.

Using the spreadsheet for plotting multiple trajectories we get the graph shown in Figure 3.4, which shows the trajectories of three battles.

d. Find the FER=1 line.

Recall that the FER=1 line in the B,R-plane is given by $R = \sqrt{\dfrac{b}{r}}B$. Thus for this particular example we get the line

$$R = \sqrt{\frac{0.10}{0.15}}B$$

$$R \approx 0.82 \cdot B.$$

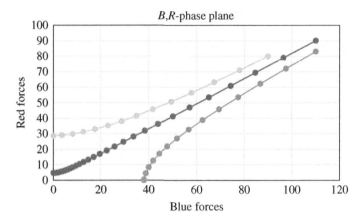

FIGURE 3.4   Excel phase plane trajectories for Exercise 3.2.1c.

e. Plot a trajectory of mutual destruction in the B,R-phase plane.

A trajectory of mutual destruction starts on the FER=1 line. Thus, if we start with $B(0) = 200$, we must start with $R = \sqrt{\dfrac{0.10}{0.15}}200 \approx 163.299$. With these initial values for Blue and Red, Figure 3.5 shows the trajectory resulting in mutual destruction.

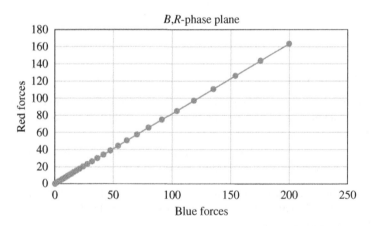

FIGURE 3.5   Mutual destruction trajectory for Exercise 3.2.1e.

3.3   THE LANCHESTER MODEL WITH REINFORCEMENTS

1   Model the course of a battle if Blue initially has 250 forces and Red has 230. Assume the fighting effectiveness for Blue is $b = 0.08$ and for Red is $r = 0.10$. At each time step 5 units of reinforcements arrive for Blue and 4 for Red.

   a. Determine by hand the numbers of Blue and Red forces remaining after two time steps.

We first use the DDS to find the numbers of forces remaining after one time step. For Blue we get

$$B(1) = B(0) - rR(0) + f$$
$$B(1) = 250 - 0.10 \cdot 230 + 5$$
$$B(1) = 232.$$

For Red we get

$$R(1) = R(0) - bB(0) + w$$
$$R(1) = 230 - 0.08 \cdot 250 + 4$$
$$R(1) = 214.$$

Now that we have the force levels after one time step, we can apply the DDS again to find the force levels after two time steps. For Blue we get

$$B(2) = B(1) - rR(1) + f$$
$$B(2) = 232 - 0.10 \cdot 214 + 5$$
$$B(2) = 215.6.$$

For Red we get

$$R(2) = R(1) - bB(1) + w$$
$$R(2) = 214 - 0.08 \cdot 232 + 4$$
$$R(2) = 199.44.$$

After two time steps Blue will have 215.6 forces remaining and Red 199.44.

b. Determine the eventual victor and how many forces remain at the end of the battle.

Here we use the Excel spreadsheet for the Lanchester model with reinforcements and drag the formulas down until one side no longer has any forces remaining. Figure 3.6 shows the Excel results with most rows hidden. Red will be the eventual victor, and it will take 52 time steps for Red to achieve victory. Note that the Excel results confirm our calculations in part a.

3 Determine the equilibrium point for the situation described in Exercise 3.3.1. Confirm the result with Excel.

The equilibrium point is given by (B^*, R^*), where

$$B^* = \frac{w}{b} = \frac{4}{0.08} = 50$$

$$R^* = \frac{f}{r} = \frac{5}{0.10} = 50.$$

	A	B	C	D	E	F
1	Lanchester Model with Reinforcements					
2						
3	Fighting Effectiveness:			Reinforcements:		
4	For Blue, b =		0.08	For Blue, f =	5	
5	For Red, r =		0.10	For Red, w =	4	
6						
7	t	B(t)	R(t)			
8	0	250	230			
9	1	232.0	214.0			
10	2	215.6	199.4			
59	51	2.5	95.5			
60	52	-2.1	99.3			

FIGURE 3.6   Excel output for Exercise 3.3.1.

Figure 3.7 shows the Excel confirmation.

	A	B	C	D	E	F
1	Lanchester Model with Reinforcements					
2						
3	Fighting Effectiveness:			Reinforcements:		
4	For Blue, b =		0.08	For Blue, f =	5	
5	For Red, r =		0.10	For Red, w =	4	
6						
7	t	B(t)	R(t)			
8	0	50	50			
9	1	50.0	50.0			
10	2	50.0	50.0			
11	3	50.0	50.0			
12	4	50.0	50.0			

FIGURE 3.7   Excel confirmation of equilibrium for Exercise 3.3.3.

5 Use Excel to determine the stability of the equilibrium point in Exercise 3.3.3.
We use the Excel worksheet to examine the course of the battle if we start off of the equilibrium point. Numerical results indicate that starting either side a little above (or a little below) its equilibrium value will result in a victory (or loss) for that side. Thus the equilibrium appears to be unstable.

7 Blue initially has 10,0000 forces and Red has 15,000. Assume the fighting effectiveness for Blue is $b = 0.08$ and for Red is $r = 0.10$. Assuming that Red has no reinforcements, determine how many reinforcements Blue needs per day in order to secure victory.

We use the Excel spreadsheet for the Lanchester model with reinforcements. We enter all known parameters and then try different values for the parameter f until we find the cut-off between a Blue defeat and a Blue victory. The result is that if Blue has at least 606 reinforcements per day then Blue will win. If Blue has 605 or fewer reinforcements per day then Red will win.

9 *Extension:* Incorporate appropriate IF statements in the Lanchester reinforcements model to turn off the model once one side reaches a force level of 0.

Once one side reaches a force level of 0 (or less), the battle is over. As a result, *both* sides should stop sending reinforcements once this occurs. Thus we have to check both force levels at each time step, and if either one of them is negative, we change the model accordingly. Because we have two requirements to check for each side, we use nested IF statements. The basic structure for Blue, for example, is, "If Blue's forces fall below 0, Blue stays at 0. Otherwise, check: if Red's forces fall below 0, then Blue stops sending reinforcements. Otherwise, if neither side falls below 0, proceed as before." The Excel command for Blue is shown in Figure 3.8.

	A	B	C	D	E	F	G	H	I	J
1	Lanchester Model with Reinforcements									
2										
3	Fighting Effectiveness:			Reinforcements:						
4	For Blue, *b* =		0.08	For Blue, *f* =		50				
5	For Red, *r* =		0.10	For Red, *w* =		20				
6										
7	*t*	*B(t)*	*R(t)*							
8	0	10000	5000							
9	1	=IF(B8-$C$5*C8+$F$4<0,0,IF(C8-$C$4*B8+$F$5<0,B8-$C$5*C8,B8-$C$5*C8+$F$4))								

FIGURE 3.8 Excel set-up for Exercise 3.3.9.

11 *Extension*: Modify the Lanchester model with reinforcements to include non-combat losses. Implement the model in Excel, and carry out an equilibrium analysis.

Assuming as we did in Exercise 3.3.10 that non-combat losses are proportional to force size, the DDS for the Lanchester model with reinforcements becomes

$$B(t) = B(t-1) - rR(t-1) + f - cB(t-1)$$

$$R(t) = R(t-1) - bB(t-1) + w - sR(t-1),$$

where $c > 0$ and $s > 0$ are new parameters reflecting the proportions of non-combat losses for each side per time step.

Implementing the model in Excel requires only a minor modification to the Lanchester spreadsheet with reinforcements. Figure 3.9 shows the set-up along with the new formula for Blue.

	A	B	C	D	E	F
1	Lanchester Model with Reinforcements and Non-combat Losses					
2						
3	Fighting Effectiveness:			Reinforcements:		
4	For Blue, b =		0.08	For Blue, f =		100
5	For Red, r =		0.10	For Red, w =		200
6						
7	Non-combat Losses:					
8	For Blue, c =		0.01			
9	For Red, s =		0.02			
10						
11	t	$B(t)$		$R(t)$		
12	0	10000		15000		
13	1	=B12-$C$5*C12+$F$4-$C$8*B12				

FIGURE 3.9   Excel set-up for Exercise 3.3.11.

Finding equilibrium points means finding points (B^*, R^*) such that

$$B^* = B^* - rR^* + f - cB^*$$

$$R^* = R^* - bB^* + w - sR^*.$$

After some simplifying we see that an equivalent system is

$$cB^* + rR^* = f$$

$$bB^* + sR^* = w.$$

Using the first equation we solve for B^* in terms of R^* to get

$$B^* = \frac{f}{c} - \frac{r}{c}R^*.$$

We substitute the above expression for B^* into the second equation to get

$$b\left(\frac{f}{c} - \frac{r}{c}R^*\right) + sR^* = w.$$

This allows us to solve for R^*. We get

$$R^* = \frac{w - \frac{bf}{c}}{s - \frac{br}{c}} = \frac{cw - bf}{cs - br}.$$

Then

$$B^* = \frac{f}{c} - \frac{r}{c}R^*$$

$$= \frac{f}{c} - \frac{r}{c}\left(\frac{cw - bf}{cs - br}\right).$$

After simplifying we find that

$$B^* = \frac{rw - sf}{rb - sc}.$$

Thus our single equilibrium point for the system is

$$(B^*, R^*) = \left(\frac{rw - sf}{rb - sc}, \frac{cw - bf}{cs - br}\right).$$

As an example, if we let $b = 0.08$, $r = 0.10$, $f = 100$, $w = 200$, $c = 0.01$, and $s = 0.02$, then we get an equilibrium point at

$$(B^*, R^*) = \left(\frac{0.10 \cdot 200 - 0.02 \cdot 100}{0.10 \cdot 0.08 - 0.02 \cdot 0.01}, \frac{0.01 \cdot 200 - 0.08 \cdot 100}{0.01 \cdot 0.02 - 0.08 \cdot 0.10}\right)$$

$$\approx (2307.7, 769.2).$$

We confirm that this is in fact an equilibrium point by plugging the values into our spreadsheet and observing that the force levels for Blue and Red do not change.

Numerical experimentation for this equilibrium point shows that if we start off of it then one side will prevail. Thus, we consider the equilibrium point unstable for these parameter choices.

3.4 HUGHES AIMED FIRE SALVO MODEL

1 Consider two fleets whose attributes are given in Text Table 3.8. Project the course of a battle between the two forces.

TEXT TABLE 3.8 Hughes Salvo Model Parameters for Exercise 3.4.1

Blue		Red	
Parameter	Value	Parameter	Value
$B(0)$	17	$R(0)$	14
b	4	r	6
c	2	s	3
d	5	u	4

a. Determine by hand the numbers of Blue and Red forces remaining after one time step.

Here we plug all parameters into the DDS for the Hughes model, and we use a calculator to find the numbers of forces after one time step:

$$B(1) = B(0) - \frac{6 \cdot R(0) - 2 \cdot B(0)}{5} = 17 - \frac{6 \cdot 14 - 2 \cdot 17}{5} = 7$$

$$R(1) = R(0) - \frac{4 \cdot B(0) - 3 \cdot R(0)}{4} = 14 - \frac{4 \cdot 17 - 3 \cdot 14}{4} = 7.5.$$

After one salvo, Blue will have 7 forces remaining and Red will have 7.5 forces remaining.

b. Determine the eventual victor and how many ships remain at the end of the battle.

We plug the given parameters into our Hughes model spreadsheet. Figure 3.10 shows that Red is the eventual winner after 3 salvos.

	A	B	C	D	E	F	G
1	The Hughes Salvo Model						
2					**Blue**		**Red**
3	Offensive firepower:			$b =$	4	$r =$	6
4	Defensive power:			$c =$	2	$s =$	3
5	Staying power:			$d =$	5	$u =$	4
6							
7	t	$B(t)$	$R(t)$				
8	0	17	14				
9	1	7.0	7.5				
10	2	0.8	6.1				
11	3	0.0	6.1				

FIGURE 3.10   Excel output for Exercise 3.4.1.

3   For the parameters given in Text Table 3.8, use the FER to determine the eventual winner.

We plug all required parameters into the formula for FER to get

$$FER = \frac{uR(0)(rR(0) - cB(0))}{dB(0)(bB(0) - sR(0))} = \frac{4 \cdot 14(6 \cdot 14 - 2 \cdot 17)}{5 \cdot 17(4 \cdot 17 - 3 \cdot 14)} \approx 1.267.$$

Since $FER > 1$, we predict a victory for Red, a prediction that is born out in Exercise 3.4.1.

5   For the parameters given in Text Table 3.8, use the fighting strength to determine the eventual winner.

The fighting strength for Blue is given by

$$dB(0)(bB(0) - sR(0)) = 5 \cdot 17(4 \cdot 17 - 3 \cdot 14) = 2,210.$$

The fighting strength for Red is given by

$$uR(0)(rR(0) - cB(0)) = 4 \cdot 14(6 \cdot 14 - 2 \cdot 17) = 2,800.$$

Since Red has the greater fighting strength, Red will be the winner.

7   Suppose we know the force attributes for Red and Blue are as given in Text Table 3.10. Suppose also that Red has the resources to increase any of its quality parameters by 1. Determine the optimal way Red should make improvements to its fleet in order to face Blue.

TEXT TABLE 3.10   Hughes Salvo Model Parameters for Exercise 3.4.7

Blue		Red	
Parameter	Value	Parameter	Value
$B(0)$	20	$R(0)$	15
b	4	r	5
c	2	s	2
d	4	u	6

With the parameters given in Text Table 3.10, Blue will defeat Red in 3 salvos. Red has the ability to improve one of its parameters: offensive firepower, defensive power, or staying power. The question of which one Red should improve can be answered via trial and error with Excel. If Red increases its offensive firepower from 5 to 6, Red will win the battle in 3 salvos with 3.9 forces remaining. If instead Red increases its defensive power from 2 to 3, Red will win the battle in 4 salvos with 5.8 forces remaining. Finally, if Red decides to increase its staying power from 6 to 7, then Blue still wins the battle, though it takes 4 salvos and costs Blue more forces. Thus the best choice for Red when facing Blue would be to increase its defensive power by 1.

9   Let our parameter choices be given by Text Table 3.12. Analyze the global behavior of the model using a phase plane diagram.

TEXT TABLE 3.12   Hughes Salvo Model Parameters for Exercise 3.4.9

Blue		Red	
Parameter	Value	Parameter	Value
b	10	r	10
c	3	s	5
d	3	u	5

As discussed in the text, there are four crucial lines through the origin that determine the fate of a battle for a given choice of $B(0)$ and $R(0)$. Those four lines are: $R = \dfrac{c}{r}B,\ R = \dfrac{c+d}{r}B,\ R = \dfrac{b}{s+u}B$, and $R = \dfrac{b}{s}B$. Our choices of parameters give us the following four slopes for our lines: $\dfrac{c}{r} = 0.30,\ \dfrac{c+d}{r} = 0.60,\ \dfrac{b}{s+u} = 1$, and $\dfrac{b}{s} = 2$. These four lines through the origin divide the phase plane into 5 distinct regions (see Figure 3.11), each corresponding to a different fate for our model.

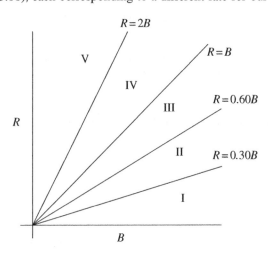

FIGURE 3.11   Phase plane diagram for Exercise 3.4.9.

Working from the bottom up we consider each region.

- Region I corresponds to points where Red will be completely eliminated in the next salvo with no harm caused to Blue.
- Region II corresponds to points where Red will be completely eliminated in the next salvo but will cause some harm to Blue.
- Region III corresponds to points where both Red and Blue will be completely eliminated in the next salvo.
- Region IV corresponds to points where Blue will be completely eliminated in the next salvo but will cause some harm to Red.
- Region V corresponds to points where Blue will be completely eliminated in the next salvo and Red will not be harmed.

Note that for the parameter choices in this exercise the battle cannot last longer than one salvo.

3.5   ARMSTRONG SALVO MODEL WITH AREA FIRE

1   Suppose Blue knows through scouting that all Red units are located somewhere within an area of 10 mile^2. Each missile that Blue fires has an area of lethality of 0.02 mile^2. Find the target area ratio for Blue.

By definition, the target area ratio is given by

$$\frac{\text{missile's area of lethality}}{\text{area enemy known to be in}} = \frac{0.02}{10} = 0.002.$$

3 Suppose we have two forces with attributes given in Text Table 3.19. Use the Hughes salvo model and the area fire salvo model to determine which type of battle Blue should prefer.

TEXT TABLE 3.19   Area Fire Model Parameters for Exercise 3.5.3

Blue		Red	
Parameter	Value	Parameter	Value
$B(0)$	220	$R(0)$	650
b	7	r	4
c	5	s	2
d	7	u	3
e	0.003	v	0.003

First we plug all parameters (except the target area ratios) into the Hughes Excel spreadsheet. As Figure 3.12 shows, in an aimed fire situation Red accomplishes a decisive victory.

	A	B	C	D	E	F	G
1	The Hughes Salvo Model						
2					**Blue**		**Red**
3	Offensive firepower:			$b =$	7	$r =$	4
4	Defensive power:			$c =$	5	$s =$	2
5	Staying power:			$d =$	7	$u =$	3
6							
7	t	$B(t)$	$R(t)$				
8	0	220	650				
9	1	5.7	570.0				
10	2	0.0	570.0				
11	3	0.0	570.0				
12	4	0.0	570.0				
13	5	0.0	570.0				

FIGURE 3.12   Excel output for Hughes model for Exercise 3.5.3.

Next we plug all parameters into the Armstrong Excel spreadsheet. As Figure 3.13 shows, in an area fire situation Blue will be the victor.

We see that Blue should prefer an area fire battle, one in which neither side knows the exact location of the other. In such a situation the superior quality

	A	B	C	D	E	F	G
1	Area Fire Salvo Model						
2					**Blue**		**Red**
3	Offensive firepower:			$b =$	7	$r =$	4
4	Defensive power:			$c =$	5	$s =$	2
5	Staying power:			$d =$	7	$u =$	3
6	Target area ratio:			$e =$	0.003	$v =$	0.003
7							
8	t	$B(t)$	$R(t)$				
9	0	220	650				
10	1	132.0	82.3				
11	2	132.0	61.1				
34	25	132.0	0.1				
35	26	132.0	0.0				

FIGURE 3.13 Excel output for Armstrong model for Exercise 3.5.3.

of the Blue forces will be enough to overcome the superior numbers of the Red forces. Red, on the other hand, will do whatever it can to engage Blue in an aimed fire battle even if it means revealing its location to Blue. In such a battle, the superior numbers of the Red forces more than make up for Red's inferior quality. Note that both sides have equal target area ratios, so the difference in preferred strategy is not due to superior scouting or missile lethality.

5 For the parameters given in Text Table 3.19 sketch the phase plane diagram that summarizes all possible results. Test each region with an Excel example.

We use the phase plane diagram from the text as our guide, and we plug in the given parameters to find the cut-offs. The result is shown in Figure 3.14.

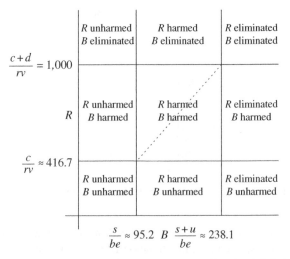

FIGURE 3.14 Phase plane diagram for Exercise 3.5.5.

Suppose we start with 240 Blue forces and 500 Red forces, which puts us in the rightmost middle region. According to our diagram, this should result in Blue being harmed (but not eliminated) and Red being eliminated in the next salvo. Plugging these initial values into the area fire Excel spreadsheet gives the results shown in Figure 3.15, which confirm our prediction based on the phase plane diagram.

	A	B	C	D	E	F	G
1	Area Fire Salvo Model						
2					**Blue**		**Red**
3	Offensive firepower:			$b =$	7	$r =$	4
4	Defensive power:			$c =$	5	$s =$	2
5	Staying power:			$d =$	7	$u =$	3
6	Target area ratio:			$e =$	0.003	$v =$	0.003
7							
8	t	$B(t)$	$R(t)$				
9	0	250	500				
10	1	214.3	0.0				
11	2	214.3	0.0				
12	3	214.3	0.0				

FIGURE 3.15   Excel output for Exercise 3.5.5.

The other regions in the diagram are tested similarly.

7 *Extension*: Modify the basic Lanchester model so that it represents an area fire situation. Carry out a full analysis of the model including numerical experiments, equilibrium analysis, and a phase plane analysis.

Recall that the basic Lanchester model is given by the DDS

$$B(t) = B(t-1) - rR(t-1)$$
$$R(t) = R(t-1) - bB(t-1).$$

If each side only knows the approximate location of the other, then as before only some proportion of possible attacks are on target. Proceeding as we did for the Armstrong model, the DDS that accounts for this is given by

$$B(t) = B(t-1) - rvR(t-1)B(t-1)$$
$$R(t) = R(t-1) - beB(t-1)R(t-1).$$

In Figure 3.16 we show the Lanchester area fire spreadsheet set-up with the equation for Blue showing. Once the spreadsheet is set-up correctly, numerical experimentation is a matter of trying different parameter values as well as different values for initial force levels.

	A	B	C	D	E	F	G
1	Lanchester Area Fire Model						
2					**Blue**		**Red**
3	Offensive firepower:			b =	0.1	r =	0.05
4	Target area ratio:			e =	0.002	v =	0.003
5							
6	t	B(t)	R(t)				
7	0	250	500				
8	1	=B7-$G$3*$G$4*C7*B7					
9	2	214.8	453.0				

FIGURE 3.16   Excel set-up for Exercise 3.5.7.

Equilibrium points are points (B^*, R^*) such that

$$B^* = B^* - rvR^*B^*$$
$$R^* = R^* - beB^*R^*.$$

This system simplifies to

$$0 = -rvR^*B^*$$
$$0 = -beB^*R^*.$$

The only solutions to the system are where either $B^* = 0$, $R^* = 0$, or both. Thus there is no possibility for a stalemate.

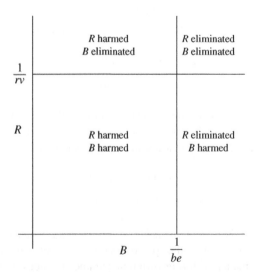

FIGURE 3.17   Phase plane diagram for Exercise 3.5.7.

To create the phase plane we first note that the model only gives physically meaningful results when the number of forces lost by a side is both positive and less than the number of forces that remain. Thus for Blue we must have

$$0 < rvB(t-1)R(t-1) < B(t-1), \text{ and for Red}$$

$$0 < beB(t-1)R(t-1) < R(t-1).$$

After some algebra we get the equivalent inequalities

$$0 < R(t-1) < \frac{1}{rv}, \text{ and} \tag{3.1}$$

$$0 < B(t-1) < \frac{1}{be}.$$

If $R(t-1) \geq \frac{1}{rv}$, Blue will be eliminated in the next time step. If $B(t-1) \geq \frac{1}{be}$, Red will be eliminated in the next time step. As long as both sides have positive force levels, each will harm the other. This leads us to the phase plane diagram given in Figure 3.17.

4

THE SPREAD OF INFECTIOUS DISEASES

4.1 THE S-I-R MODEL

1 Given $R(0) = 0$, $N = 5{,}000$, $I(0) = 5$, $\beta = 0.9$, and $\delta = 5$, produce a graph of the S-I-R model over time, and use it to answer the following questions.

a. Approximately when was the epidemic at its worst?

The graph of the model is given in Figure 4.1. The epidemic is at its worst when the graph for the number of infectives is at its peak. By inspection this peak occurs at about day 14. We can confirm this result by checking the Excel numerical output and noting that the highest value in the infective column occurs at day 14.

b. How many people were sick at the peak of the epidemic?

Again by inspection of the graph we see that about 2,500 people were sick at the epidemic's peak. By looking at the Excel numerical output we can be more precise: on day 14 there were approximately 2,466 infectives.

c. Approximately how long did the epidemic last?

The epidemic lasts until the number of infectives is roughly 0. From the graph it is difficult to be precise, but sometime around day 50 the number of infectives appears to be about 0. By looking at the Excel output rather than the graph we can get a better estimate.

Solutions Manual to Accompany Models for Life: An Introduction to Discrete Mathematical Modeling with Microsoft® Office Excel®, First Edition. Jeffrey T. Barton.
© 2016 John Wiley & Sons, Inc. Published 2016 by John Wiley & Sons, Inc.
Companion website: www.wiley.com/go/barton/solutionsmanual_modelsforlife

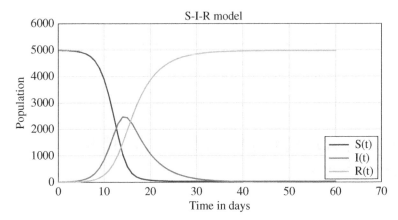

FIGURE 4.1   Excel graph for Exercise 4.1.1.

d. Approximately how many people contracted the disease during the course of the epidemic?

Here we wait until the epidemic is over, then we see how many people ended up in the removed category since anyone in the removed category must have first been ill. According to the graph, nearly everyone will have been ill at the end of the epidemic. By looking at the Excel numerical output rather than the graph we can be more precise: about 4,975 people out of the 5,000 total will have been ill by the time the epidemic is over.

3   Suppose that during the Eyam plague the duration of infectivity was 10 days rather than 11 so that $\delta = 10$. Graphically determine the corresponding best choice for β to two decimal places of accuracy.

We use the S-I-R spreadsheet with the Eyam data included, change the parameter δ to 10, and use trial and error to graphically determine the optimum choice for β. Figure 4.2 shows the graph for $\beta = 0.16$, which appears to be the choice that provides the best-fitting S-I-R graph to the data.

5   Suppose an epidemic has $\beta = 2$, and $\delta = 6$.

a. Determine the maximum number of new infections a single infective would be expected to cause.

We expect a maximum number of new infections caused by a single infective to occur when everyone else is still susceptible to the disease. In other words, we need to find the basic reproductive rate:

$$R_0 = \beta \times \delta = 2 \times 6 = 12.$$

b. Explain why we do not typically expect a single infective to cause that many new cases.

Typically a population is not wholly susceptible – some members are already infective and some members have already recovered. Thus some of the

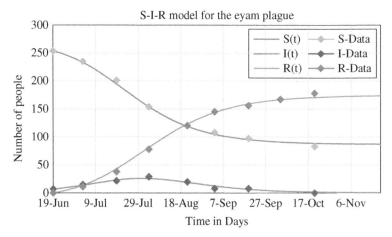

FIGURE 4.2   Excel graph for Exercise 4.1.3.

effective contacts had by an infective will not result in new infectives, and this keeps the number of new cases caused by a typical infective below the basic reproductive rate.

7 *Extension*: Use an IF statement to modify the Excel S-I-R model so that it will not report negative values of susceptibles. (See Chapter 3 for similar examples involving the Lanchester model and the Hughes model.)

We know that the number of new infectives on any given day cannot exceed the number of susceptibles from the previous day. Thus to make sure our S-I-R model only reports physically meaningful numbers, we want to set the number of susceptibles to 0 whenever the number of new infectives exceeds the previous number of susceptibles. In other words, if

$$\beta I(t-1)\frac{S(t-1)}{N} > S(t-1),$$

then we set $S(t)=0$. Similarly, if

$$\beta I(t-1)\frac{S(t-1)}{N} > S(t-1),$$

then we only add $S(t-1)$ to the infective category. Thus we would have

$$I(t) = I(t-1) + S(t-1) - \frac{1}{\delta}I(t-1)$$

in such a case. Figure 4.3 shows the Excel set-up with the formula for the infectives category displayed.

	A	B	C	D	E	F	G	H	I	J	K
1	S-I-R Model										
2											
3	Effective contact rate, β =			10							
4	Duration of infectivity, δ =			2							
5	Total population, N =			3000							
6											
7	t	S(t)	I(t)	R(t)							
8	0	2990	10	0							
9	1	2890.3	=IF($D$3*C8*B8/$D$5>B8,C8+B8-(1/$D$4)*C8,C8+$D$3*C8*B8/$D$5-(1/$D$4)*C8)								

FIGURE 4.3   Excel set-up for Exercise 4.1.7.

9   Repeat Exercise 4.1.8 with a different epidemic.

a. For example purposes we use the 1918 measles epidemic from Kansas. The duration of infectivity is given as 6-7 days, so we use $\delta = 6.5$. We use the basic reproductive rate to find β:

$$R_0 = \beta \times \delta$$
$$5.5 = \beta \times 6.5$$
$$0.846 \approx \beta.$$

b. Produce a graph for the epidemic over a suitable time period.

According to Demographia.com, the population of Kansas in 1918 was about 1.7 million. Assuming 10 people were initially infected, we use the basic S-I-R spreadsheet to produce the graph in Figure 4.4, which projects the epidemic over a period of about 4 months.

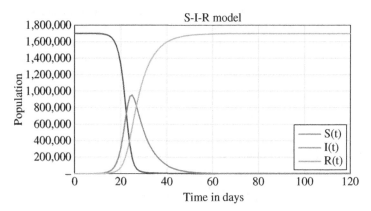

FIGURE 4.4   Excel graph for Exercises 4.1.8 and 4.1.9.

c. Apply the Threshold Theorem to determine how many susceptibles were present when the epidemic began to die out.

The number of susceptibles at which the epidemic begins to wane is given by

$$S = \frac{N}{R_0} = \frac{1,700,000}{5.5} \approx 309,091.$$

d. Verify the theorem's result by noting the appropriate point on your graph.

The peak of the epidemic occurs at about day 25. We then estimate the number of susceptibles on that day, and it appears to be about 300,000 on the graph.

e. Determine how many people were sick at the peak of the epidemic.

By examining Excel's numerical output we see that the peak occurred on day 25 with about 951,136 people ill.

f. Determine how many people total became ill during the epidemic.

The total number who became ill will be the total number in the removed category after the epidemic is over, about 1,697,639.

g. Determine the vaccination percentage required to prevent an epidemic of this disease.

We apply the Herd Immunity Theorem to determine that we would need to vaccinate a percentage equal to

$$1 - \frac{1}{R_0} = 1 - \frac{1}{5.5} \approx .818 = 81.8\%.$$

11 *Extension*: An important consideration when proposing a vaccination strategy is the vaccine's **efficacy**, i.e. the percentage of vaccines administered that actually achieve the desired immunity. Incorporate a parameter for vaccine efficacy into the modified S-I-R model in Exercise 4.1.10.

Only vaccines that work have the effect of moving susceptibles to the removed category. Thus if we let ε be the percentage of vaccinations that actually achieve immunity, then the number of susceptibles who are initially moved to the removed category becomes

$$\varepsilon \rho N.$$

Figure 4.5 shows the Excel set-up with the formula for susceptibles displayed.

	A	B	C	D
1	S-I-R Model			
2				
3	Effective contact rate, β =			0.846
4	Duration of infectivity, δ =			6.5
5	Total population, N =			1,700,000
6	Vaccination percentage, ρ =			81.8%
7	Vaccine efficacy percentage, ε =			90.0%
8				
9	t	$S(t)$	$I(t)$	$R(t)$
10	0	=D5-C10-D7*D6*D5		1,251,540

FIGURE 4.5 Excel set-up for Exercise 4.1.11.

13 Is it possible to determine the total number of people who become ill by looking only at the Infectives column in the S-I-R model spreadsheet? Explain.

Yes, it is possible, though we have to be careful. We cannot simply sum all of the entries in the infectives column. That would produce an overestimate because people spend more than one day in the infective category – we would be counting the same person multiple times. However, if we sum all of the entries in the infectives column and then divide by the average duration of infectivity, we should arrive at the total number who become ill.

4.2 S-I-R WITH VITAL DYNAMICS

1 Consider an epidemic of mumps in a city of 1,000,000 similar to the one in Baltimore, Maryland, in 1943. Assume that initially 10 people are infective.

    a. Use the S-I-R model with vital dynamics to model the epidemic; from Text Table 4.2 use $\delta = 8$, $R_0 = 7$, and $\mu = 0.0006$.

TEXT TABLE 4.2   Estimated Values for R_0 and δ for Several Historical Epidemics

Infection	Location	Time period	R_0	δ (days)
Measles	Cirencester, England	1947–50	13–14	6–7
	England and Wales	1950–68	16–18	
	Kansas, U.S.	1918–21	5–6	
	Ontario, Canada	1912–13	11–12	
	Willesden, England	1912–13	11–12	
	Ghana	1960–68	14–15	
	Eastern Nigeria	1960–68	16–17	
Pertussis	England and Wales	1944–68	16–18	7–10
	Maryland, U.S.	1943	16–17	
	Ontario, Canada	1912–13	10–11	
Chicken Pox	Maryland, U.S.	1913–17	7–8	10–11
	New Jersey, U.S.	1912–21	7–8	
	Baltimore, U.S.	1943	10–11	
	England and Wales	1944–68	10–12	
Diphtheria	New York, U.S.	1918–19	4–5	2–5
	Maryland, U.S.	1908–17	4–5	
Scarlet fever	Maryland, U.S.	1908–17	7–8	14–21
	New York, U.S.	1918–19	5–6	
	Pennsylvania, U.S.	1910–16	6–7	
Mumps	Baltimore, U.S.	1943	7–8	4–8
	England and Wales	1960–80	11–14	
	Netherlands	1970–80	11–14	
Rubella	England and Wales	1960–70	6–7	11–12
	West Germany	1970–7	6–7	
	Czechoslovakia	1970–7	8–9	
	Poland	1970–7	11–12	
	Gambia	1976	15–16	

TEXT TABLE 4.2 (*continued*)

Infection	Location	Time period	R_0	δ (days)
Poliomyelitis	U.S.	1955	5–6	14–20
	Netherlands	1960	6–7	
HIV (Type 1)	England and Wales (male homosexuals)	1981–5	2–5	
	Nairobi, Kenya (female prostitutes)	1981–5	11–12	
	Kampala, Uganda (heterosexuals)	1985–7	10–11	

Source: Anderson and May 1991, Table 4.1, p. 70. Reproduced with permission of Oxford University Press.

Before we can use our Excel spreadsheet for the S-I-R model with vital dynamics, we must find β. Since we already have $R_0 = 7$, we find β as follows:

$$R_0 = \frac{\beta\delta}{1 + \mu\delta}$$

$$7 = \frac{\beta \cdot 8}{1 + 0.0006 \cdot 8}$$

$$7.0336 = \beta \cdot 8$$

$$0.879 \approx \beta.$$

Now we plug all parameters into our Excel spreadsheet. The set-up is shown in Figure 4.6.

b. Graph the epidemic over a period of 1 year. Compare your graph to the usual graph from the S-I-R model. How is the new graph different? What accounts for the difference?

	A	B	C	D
1	Basic S-I-R Model with Vital Dynamics			
2				
3	Effective contact rate, $\beta =$			0.879
4	Duration of infectivity, $\delta =$			8
5	Total population, $N =$			1,000,000
6	Birth and death rate, $\mu =$			0.0006
7				
8	t	$S(t)$	$I(t)$	$R(t)$
9	0	999,990	10	0
10	1	999,981.2	17.5	1.3
11	2	999,965.8	30.7	3.4
12	3	999,938.8	53.9	7.3
13	4	999,891.5	94.5	14.0
14	5	999,808.5	165.7	25.8
15	6	999,662.9	290.5	46.5

FIGURE 4.6 Excel set-up for Exercise 4.2.1.

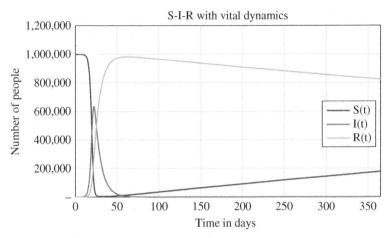

FIGURE 4.7   Excel graph for Exercise 4.2.1b.

The graph over 1 year is shown in Figure 4.7. This graph is different from the original S-I-R graph because the number in the removed category eventually starts to decrease over time while the number in the susceptible category eventually starts to increase over time. Both differences are due to the presence of vital dynamics: people are leaving the removed category by dying from causes other than the disease, and newborns are being added to the susceptible category.

c. Graph the epidemic over a period of 2 years. What new phenomenon do you observe? Explain why it is happening.

Figure 4.8 shows the 2-year graph. The graph shows a second epidemic wave occurring around day 560. The reason for the second epidemic wave is that the susceptible category has been replenished due to new births. Once

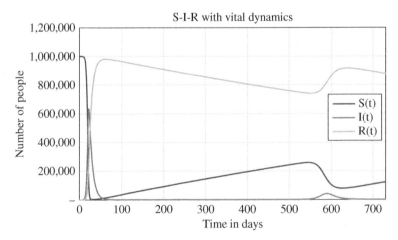

FIGURE 4.8   Excel graph for Exercise 4.2.1c.

the number of susceptibles is large enough, an epidemic is once again able to occur.

d. Graph the epidemic over a period of 10 years. Describe what you see, paying particular attention to the differences between your new graph and the standard S-I-R graph.

Figure 4.9 shows the 10-year graph. Unlike a basic S-I-R graph, Figure 4.9 shows several epidemic waves occurring. The numbers in each category eventually level off at some positive value.

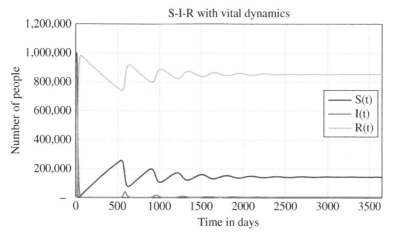

FIGURE 4.9   Excel graph for Exercise 4.2.1d.

e. What does Excel predict for the long-term number of infectives? What does this mean practically?

By holding our cursor over the point at which the number of infectives has leveled off, we see that long-term there will be about 4133 infectives in our population. This means that the disease will be in an endemic state, where it is always present at some low level and is never entirely eradicated.

3 Repeat Exercise 4.2.1 for an epidemic of your choosing. Continue to use the birth and death rate of $\mu = 0.0006$.

The only difference between this exercise and Exercise 4.2.1 is that the parameter values will change depending on the epidemic selected.

5 *Extension*: Modify the S-I-R with vital dynamics Excel spreadsheet so that it automatically calculates the non-trivial equilibrium values for an epidemic.

Here we use space at the top of the spreadsheet to include the necessary calculations. Because it is used in all of the formulas, we also include a cell where we store the basic reproductive number. The set-up is given in Figure 4.10 with

	A	B	C	D	E	F	G	H
1	Basic S-I-R Model with Vital Dynamics							
2								
3	Effective contact rate, $\beta =$			0.879		$R_0 =$	7	
4	Duration of infectivity, $\delta =$			8		$S^* =$	142,857.14	
5	Total population, $N =$			1,000,000		$I^* =$	=(D6*D5/D3)*(G3-1)	
6	Birth and death rate, $\mu =$			0.0006		$R^* =$	853,242.32	
7								
8	t	$S(t)$	$I(t)$	$R(t)$				
9	0	999,990	10	0				
10	1	999,981.2	17.5	1.3				

FIGURE 4.10 Excel set-up for Exercise 4.2.5.

the formula for I^* displayed. Now whenever we change parameter values, the long-term values for the model will be found for us automatically.

4.3 DETERMINING PARAMETERS FROM REAL DATA

1 Suppose a mysterious flu-like illness is spreading on a college campus of 20,000 students. Originally only 1 student was infective, but by the time school officials became concerned 1 week later there were 50 cases. If the duration of infectivity for the disease is known to be 5 days and $\mu = 0.0005$, determine R_0 for the disease.

Assuming that the number of infectives grows exponentially early in an epidemic, we use the approximation $I(t) \approx (1+r)^t I(0)$. The information in the problem allows us to solve for r:

$$I(7) = (1+r)^7 \cdot 1$$
$$50 = (1+r)^7$$
$$50^{\frac{1}{7}} = 1 + r$$
$$0.749 \approx r.$$

Once we have r we use the relationship $R_0 = 1 + \delta r$. We get

$$R_0 = 1 + 5 \cdot 0.749 = 4.745.$$

3 Consider a disease like measles that has a very high R_0. Using Text Table 4.2, in this case assume that $R_0 = 15$ and $\delta = 7$. Estimate the growth rate of the number of infectives at the beginning of a measles epidemic.

Using the relationship $R_0 = 1 + \delta r$, we can solve for r:

$$15 = 1 + 7r$$
$$14 = 7r$$
$$2 = r.$$

5 Continuing Exercise 4.3.4, determine the effective contact rate, β, for pertussis in this population. Use Text Table 4.2 to estimate the duration of infectivity.

Text Table 4.2 allows us to estimate $\delta = 8.5$, though any value in the given range of 7–10 days is fine. Since we have $R_0 = 13.89$ from Exercise 4.3.4, we use the formula $R_0 = \dfrac{\beta\delta}{1+\mu\delta}$ to find β:

$$13.89 = \frac{\beta \cdot 8.5}{1 + 0.0005 \cdot 8.5}$$

$$13.949 \approx \beta \cdot 8.5$$

$$1.641 \approx \beta.$$

7 *Extension*: One reason the S-I-R model with vital dynamics is not the most appropriate model for Ebola is that EVD has a considerable incubation period (approximately 10 days) during which time the person who has the disease is asymptomatic and cannot transmit it. A better choice of model would be an S-E-I-R model with vital dynamics, where we add a category for those who are Exposed but not yet infectious.

a. Draw a careful flow diagram for an S-E-I-R model with vital dynamics.

Only infectives can spread the disease, but now new cases go into the Exposed category while the disease incubates. We introduce a new parameter, η, for the incubation period in days. By the Waiting Time Principle, we know that the fraction leaving the Exposed category each day is therefore $\frac{1}{\eta}$.

The flow diagram for the new model is given in Figure 4.11. Note that those in the Exposed category are also subject to the death rate from causes other than the disease.

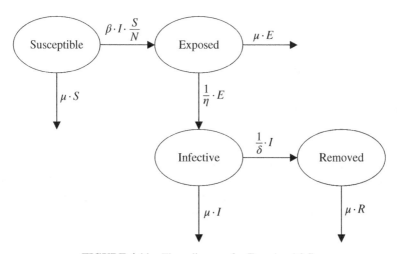

FIGURE 4.11 Flow diagram for Exercise 4.3.7.

b. Give the DDS for the model.

The DDS is given by

$$S(t) = S(t-1) - \frac{\beta I(t-1)}{N} S(t-1) - \mu S(t-1) + \mu N$$

$$E(t) = E(t-1) + \frac{\beta I(t-1)}{N} S(t-1) - \frac{1}{\eta} E(t-1) - \mu E(t-1)$$

$$I(t) = I(t-1) + \frac{1}{\eta} E(t-1) - \frac{1}{\delta} I(t-1) - \mu I(t-1)$$

$$R(t) = R(t-1) + \frac{1}{\delta} I(t-1) - \mu R(t-1).$$

c. Implement the model in Excel.

We need an additional column for the exposed category, and we need a cell in which to store the new parameter for the incubation period. The set-up is given in Figure 4.12 with the formula for the exposed category displayed.

▲	A	B	C	D	E	F	C
1	S-E-I-R Model with Vital Dynamics						
2							
3	Effective contact rate, β =				0.879		
4	Duration of infectivity, δ =				8		
5	Total population, N =				1,000,000		
6	Birth and death rate, μ =				0.0006		
7	Incubation period, η =				10		
8							
9	t	$S(t)$	$E(t)$	$I(t)$	$R(t)$		
10	0	999,990	0	10	0		
11	1	999,981.2	=C10+$E$3*D10*B10/$E$5-(1/$E$7)*C10-$E$6*C10				
12	2	999,973.5	15.6	8.5	2.3		

FIGURE 4.12 Excel set-up for Exercise 4.3.7.

d. How does the course of the potential epidemic in Dallas change when modeled by the S-E-I-R model with vital dynamics?

Here we plug all parameters found in the text into the model, including 10 days for the new incubation parameter. The peak of the epidemic is significantly delayed. In the S-E-I-R model the peak occurs around day 870 with about 8,050 infectives and an additional 13,400 that have been exposed but are not yet infective. The entire epidemic is now projected to last around 1,500 days.

4.4 S-I-R WITH VITAL DYNAMICS AND ROUTINE VACCINATIONS

1 According to the World Health Organization, vaccination coverage for measles in the United States was 91.9% in 2014. Using values for R_0 from Text Table 4.2 and life expectancy estimates from an Internet source, estimate the average age at infection for measles in the United States.

The R_0 values for measles tend to be high. Here we use $R_0 = 14$. According to CDC.gov, life expectancy in the U.S. is about 79 years. Thus the average age at infection for measles is given by

$$A = \frac{L}{R_0(1-\rho)} = \frac{79}{14 \cdot (1-.919)} \approx 69.7.$$

With such high vaccination coverage, the average age at infection is also very high at almost 70 years.

3 Our minimum vaccination coverage required to prevent an epidemic has been based on the Herd Immunity Theorem. A second way to think of disease eradication would be to vaccinate enough people so that the long-term value for the number of infectives equals zero. In other words, we could vaccinate so that $I^* = 0$. Show that the vaccination level implied by this requirement turns out to be the same as that given in the Herd Immunity Theorem.

From the text we know that the long-term number of infectives for the S-I-R model with vital dynamics and ongoing vaccinations of newborns is given by

$$I^* = \frac{\mu N}{\beta}(R_0(1-\rho)-1).$$

Setting equal to zero and solving for ρ gives us

$$0 = \frac{\mu N}{\beta}(R_0(1-\rho)-1)$$

$$0 = (R_0(1-\rho)-1)$$

$$0 = R_0 - R_0 \cdot \rho - 1$$

$$R_0 \cdot \rho = R_0 - 1$$

$$\rho = \frac{R_0 - 1}{R_0} = 1 - \frac{1}{R_0}.$$

This is the same vaccination coverage given in the Herd Immunity Theorem.

5 *Extension*: A reason that we do not actually vaccinate newborns in practice is that newborns are temporarily protected from many diseases by antibodies that are passed on from the mother. This phenomenon is known as **maternal antibody protection**, and if a vaccine is administered too soon after birth maternal antibodies can interfere with the vaccine working properly. Suppose that maternal antibody protection lasts 90 days. Modify the S-I-R model with vital dynamics and routine vaccinations to include a compartment, *M*, for maternal antibody protection.

a. Give a flow diagram for the M-S-I-R model with vital dynamics and routine vaccinations.

Newborns are now added to the new category *M* where they are protected for a period of time by maternal antibodies. We introduce a new parameter, τ, for the duration of maternal antibody protection, and in this example $\tau = 90$. With the M-S-I-R model we no longer vaccinate newborns. Instead, we wait for the maternal antibody protection to expire and then vaccinate infants as they transition from *M* to *S*. By the Waiting Time Principle, the fraction of infants moving from *M* to *S* each day is given by $\frac{1}{\tau}$. Those in the new category still experience the effects of vital dynamics. Figure 4.13 gives the flow diagram for the new model.

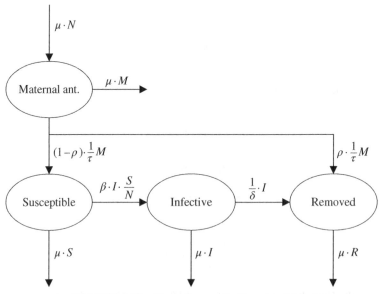

FIGURE 4.13   Flow diagram for Exercise 4.4.5.

	A	B	C	D	E	F
1	M-S-I-R with Vital Dynamics and Ongoing Vaccinations					
2						
3	Effective contact rate, β =				0.9	
4	Duration of infectivity, δ =				10	
5	Total population, N =				5000	
6	Birth and death rate, μ =				0.0005	
7	Vaccination proportion, ρ =				0.89	
8	Maternal antibody protections, τ =				90	
9						
10	t	$M(t)$	$S(t)$	$I(t)$	$R(t)$	
11	0	0	4995	5	0	
12	1	=B11+$E$6*$E$5-(1/$E$8)*B11-$E$6*B11				
13	2	4.97097	4977.4	16.2	1.4	
14	3	7.41325	4960.5	29.0	3.1	
15	4	9.82718	4932.1	52.0	6.1	
16	5	12.2131	4883.5	93.0	11.4	

FIGURE 4.14 Excel set-up for Exercise 4.4.5.

b. Give the DDS for the model.

The DDS is given by

$$M(t) = M(t-1) + \mu N - \frac{1}{\tau}M(t-1) - \mu M(t-1)$$

$$S(t) = S(t-1) - \frac{\beta I(t-1)}{N}S(t-1) - \mu S(t-1) + (1-\rho)\frac{1}{\tau}M(t-1)$$

$$I(t) = I(t-1) + \frac{\beta I(t-1)}{N}S(t-1) - \frac{1}{\delta}I(t-1) - \mu I(t-1)$$

$$R(t) = R(t-1) + \frac{1}{\delta}I(t-1) - \mu R(t-1) + \rho\frac{1}{\tau}M(t-1).$$

c. Implement the model in Excel.

In Figure 4.14 we show the set-up of the Excel model with the formula for the M compartment displayed.

5

DENSITY DEPENDENT POPULATION MODELS

5.1 THE DISCRETE LOGISTIC MODEL

1 Recall that for baleen whales we assume that $r = 5\%$, and $K = 400,000$. If in 1985 the population was $P(0) = 75,000$, determine how long it will take for the population to reach 300,000 blue whale units.

We plug all parameters into the Excel spreadsheet for the discrete logistic model and copy the formulas down until the population reaches or exceeds 300,000 BWU. The result is shown in Figure 5.1 with most rows hidden. It will take 52 years for the population to reach 300,000 BWU.

	A	B	C
1	Discrete Logistic Model		
2			
3	Intrinsic growth rate, $r =$		5.0%
4	Carrying capacity, $K =$		400,000
5			
6	t	Population	
7	0	75,000	
8	1	78,047	
58	51	298,450	
59	52	302,238	

FIGURE 5.1 Excel output for Exercise 5.1.1.

Solutions Manual to Accompany Models for Life: An Introduction to Discrete Mathematical Modeling with Microsoft® Office Excel®, First Edition. Jeffrey T. Barton.
© 2016 John Wiley & Sons, Inc. Published 2016 by John Wiley & Sons, Inc.
Companion website: www.wiley.com/go/barton/solutionsmanual_modelsforlife

3   Reveal a shortcoming of the discrete logistic model by choosing a very large initial population.

In Figure 5.2 we show the Excel output for a discrete logistic model with an intrinsic growth rate of 5%, and carrying capacity of 400,000, and an initial population of 10,000,000. The Excel output shows that in this case we would get negative population values, which are not physically meaningful. What is happening is that the growth rate formula for the discrete logistic model allows the growth rate to be less than -1 for large enough populations. I.e. if the population is large enough, the discrete logistic model can produce a decrease in the population that leads to negative values. This shortcoming of the discrete logistic model is rectified in section 5.5 with the Ricker model.

	A	B	C
1	Discrete Logistic Model		
2			
3	Intrinsic growth rate, r =		5.0%
4	Carrying capacity, K =		400,000
5			
6	t	Population	
7	0	10,000,000	
8	1	(2,000,000.00)	
9	2	(2,600,000)	
10	3	(3,575,000)	
11	4	(5,351,328)	

FIGURE 5.2   Excel output for Exercise 5.1.3.

5   *Extension*: For the general discrete logistic model find the population that results in the largest absolute population growth during the next time step.

The absolute change in the population each time step is given by the growth rate multiplied by the previous population. Thus the absolute change for the discrete logistic model is given by

$$r\left(1-\frac{P(t-1)}{K}\right)P(t-1).$$

Since it is the population that interests us we simplify the notation by suppressing the dependence on t. This gives us the expression

$$r\left(1-\frac{P}{K}\right)P.$$

This gives us a downward pointing parabola in P, and its roots are at $P=0$ and $P=K$. The vertex of the parabola occurs at the midpoint between the two roots, or $P = \dfrac{K}{2}$. Thus the maximum absolute population growth for the logistic model occurs when the population is at half of its carrying capacity.

7 Explain in a complete sentence or two why the values you found in Exercise 5.1.6 make sense intuitively.

If the population is at 0 then there is no way for the population to reproduce and it will stay at 0. If the population is at its carrying capacity, then the environment cannot support anymore. The population will not be able to grow and will remain constant.

9 Find estimates for the intrinsic growth rate and carrying capacity of the earth's human population. Be sure to cite your source(s). Use the current world population along with the parameters you found to predict the world's population over the next 50 years. Display your predictions on a graph.

There are many different estimates for the intrinsic growth rate and carrying capacity of humans. In *Dynamical Models in Biology* by M. Farkas (Farkas, 2001), the intrinsic growth rate of the human population over the last 150 years is estimated to be about 2% and the Earth's carrying capacity for humans about 12,000,000,000. The current world population is about 7,300,000,000. Figure 5.3 shows the graph of our human population model over the next 50 years.

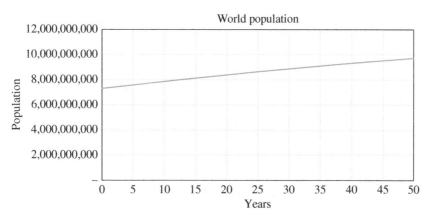

FIGURE 5.3   Excel graph for Exercise 5.1.9.

5.2   LOGISTIC GROWTH WITH ALLEE EFFECTS

1 The most recent survey of the panda population yielded an estimate of 1,864 pandas in 2014. Compare this estimate to what our model from Example 5.8 projects.

Example 5.8 uses $r = 0.92\%$, $K = 16,000$, and $S = 480$ for the parameters in the Allee effects model. The initial population of pandas was estimated to be 1,100 in 1977. To project the population in 2014, we plug all parameters into the Excel

	A	B	C
1	Allee Effects Model		
2			
3	Intrinsic growth rate, $r =$		0.92%
4	Carrying capacity, $K =$		16,000
5	Sustain. thresh., $S =$		480
6			
7	t	Population	
8	0	1100	
9	1	1112.2	
44	36	1979.2	
45	37	2029.1	

FIGURE 5.4   Excel output for Exercise 5.2.1.

Allee effects spreadsheet and drag the formulas down to $t = 37$. The results are shown in Figure 5.4 with most rows hidden. The model projects 2,029 pandas for the year 2014. Compared to the most recent population estimate of 1,864, the model projection is too high; however, the discrepancy is not that large. The model projection is about 9% higher than the most recent estimate.

3 Suppose the giant panda population did not experience Allee effects and was instead well-modeled by the discrete logistic model.

  a. Using the same parameters as in Example 5.8, project the giant panda population in 2020 using the discrete logistic model.

  Example 5.8 uses $r = 0.92\%$, $K = 16,000$, and $S = 480$ for the parameters in the Allee effects model. The initial population of pandas was estimated to be 1,100 in 1977. Without Allee effects the population model for the giant panda becomes the discrete logistic model:

  $$P(t) = P(t-1) + 0.0092\left(1 - \frac{P(t-1)}{16,000}\right)P(t-1).$$

  We plug all relevant parameters into the discrete logistic Excel spreadsheet and copy the formulas down to year 43. The result is given in Figure 5.5. Without Allee effects we would project a giant panda population of 1,579 in the year 2020.

  b. Compare the projection in a. to the projection for 2020 where Allee effects are incorporated.

  Here we use the Allee effects Excel spreadsheet to project the population to the year 43. The result is shown in Figure 5.6. The Allee effects model projects a population of about 2,395 giant pandas.

	A	B	C
1	Discrete Logistic Model		
2			
3	Intrinsic growth rate, r =		0.92%
4	Carrying capacity, K =		16,000
5			
6	t	Population	
7	0	1,100	
8	1	1,109	
49	42	1,566	
50	43	1,579	

FIGURE 5.5   Excel output for Exercise 5.2.3a.

	A	B	C
1	Allee Effects Model		
2			
3	Intrinsic growth rate, r =		0.92%
4	Carrying capacity, K =		16,000
5	Sustain. thresh., S =		480
6			
7	t	Population	
8	0	1,100	
9	1	1,112.2	
50	42	2,324.5	
51	43	2,394.7	

FIGURE 5.6   Excel output for Exercise 5.2.3b.

c. Determine by examining the Allee effects DDS what, specifically, accounts for the difference.

The factor in the Allee effects model that accounts for the sustainability threshold is given by

$$\left(\frac{P(t-1)}{S} - 1\right).$$

If the population is large relative to the sustainability threshold, S, then the Allee effects factor dramatically increases the growth rate of the model. Suppose, for example, that the giant panda population is 1,920. Then the Allee effects factor becomes

$$\left(\frac{1,920}{480} - 1\right) = 3.$$

Thus the Allee effects factor has the effect here of tripling the growth rate that would be produced by the discrete logistic model.

5 Reveal a shortcoming in the Allee effects model for the giant panda population by finding populations that will result in a negative population projection for the next time step.

This is a trial and error exercise. Initial populations above about 39,000 result in negative population projections for the following time step.
    The problem with the model is that for populations that are very large compared to the carrying capacity, the model produces growth rates that are less than -1.

7 Analyze the graph of the growth rate function for the logistic model with Allee effects. Why is this graph not realistic for some population values?

The growth rate function is the quadratic in P given by

$$r\left(1-\frac{P}{K}\right)\left(\frac{P}{S}-1\right).$$

The function is a downward pointing parabola in P. As P increases beyond the carrying capacity, the growth rate continues to become more negative, eventually dropping below -1. This is not a realistic graph for the growth rate. Once the growth rate falls below -1, the model produces negative population projections, and this does not make physical sense.

5.3 LOGISTIC GROWTH WITH HARVESTING

1 Suppose a fishery has an intrinsic growth rate of $r = 0.10$, and a carrying capacity of $K = 1,000,000$.
    a. Find the maximum sustainable yield for the fishery.

We know from the discussion in the text that the MSY is given by

$$MSY = \frac{rK}{4} = \frac{0.10 \cdot 1,000,000}{4} = 25,000.$$

    b. Assume that a constant take strategy is used to harvest fish at the MSY. What happens to the population over time if it starts at carrying capacity?

As noted in the text, using constant take harvesting with a harvesting level set at the MSY results in a semi-stable equilibrium at half of the carrying capacity. Thus if the population starts at the carrying capacity, it will decrease towards half of that value. In this example, starting at 1,000,000 will result in the population tending to the value 500,000.
    This result can be confirmed using the constant take harvesting Excel spreadsheet. If we plug in the given parameters and set the harvesting level to the value found in a. we see the population decreasing towards 500,000.

c. Assume that a constant effort strategy is used to harvest fish at the MSY. What happens to the population over time if it starts at carrying capacity?

With constant effort harvesting at the MSY we again have an equilibrium value at $\frac{K}{2} = 500,000$. This equilibrium is stable, and a population starting at 1,000,000 will therefore tend towards the value of 500,000 over time.

This result can be confirmed using the constant effort harvesting Excel spreadsheet. If we plug in the given parameters and set the harvesting level to the value found in a. we see the population decreasing towards 500,000.

3 For the fishery described in Exercise 5.3.1, answer the following.

a. Assume that a constant take strategy is used to harvest fish at the MSY. What happens to the population over time if it starts at 400,000?

Because harvesting at the MSY introduces a semi-stable equilibrium value into the constant take harvesting model, the population will crash if it falls below that equilibrium value. In this case the equilibrium value when harvesting at the MSY is 500,000. Thus if a population starts at 400,000 the population will crash.

b. Assume that a constant effort strategy is used to harvest fish at the MSY. What happens to the population over time if it starts at 400,000?

Unlike the constant take model, the constant effort model has a stable equilibrium value at 500,000 rather than a semi-stable equilibrium value. Thus if the population starts below 500,000 it will increase back towards it over time. That will be the case here for a population that starts at 400,000.

5 *Extension*: For the fishery described in Exercise 5.3.1, the MSY was set using the assumed carrying capacity of $K = 1,000,000$. Suppose a mistake was made when estimating the carrying capacity and that it actually turns out to be $K = 800,000$. Investigate the consequences of this mistake for both the constant take and constant effort models if the MSY harvest levels were set using the mistaken value of $K = 1,000,000$.

For the constant take model an intrinsic growth rate of 10% along with a carrying capacity of 800,000 results in an MSY of 20,000. Instead, due to the mistake in estimating the carrying capacity, the harvesting level was set at 25,000. Because this is more than the actual MSY, the fishery will have no equilibrium value and will crash regardless of the initial population. Evidence to support this claim can be obtained numerically by experimenting with the constant take Excel spreadsheet.

For the constant effort model our mistake is not nearly so dire. With an intrinsic growth rate of 10%, the MSY effort level is given by $\frac{10\%}{2} = 5\%$. Had the carrying capacity really been 1,000,000 then this effort level would result in an equilibrium value of $P^* = K\left(1 - \frac{e}{r}\right) = 1,000,000\left(1 - \frac{0.05}{0.10}\right) = 500,000$. The resulting harvest would be $eP^* = 0.05 \cdot 500,000 = 25,000$ fish per time step. The fact that the carrying capacity is actually only 800,000 results in an actual equilibrium value of $P^* = K\left(1-\dfrac{e}{r}\right) = 800,000\left(1-\dfrac{0.05}{0.10}\right) = 400,000$, and a resulting harvest of $eP^* = 0.05 \cdot 400,000 = 20,000$ fish per time step.

Thus in the constant effort situation our mistake in estimating the carrying capacity has caused our harvest expectation to be 5,000 higher than we actually achieve, but this is far better than in the constant effort case where our mistake would wipe out the entire population.

7 For the constant take harvest model show that if there is a semi-stable equilibrium value then it occurs at a population equal to half the carrying capacity.

A semi-stable equilibrium in the constant take model suggests that the harvest level has been set at the MSY, i.e. we have $h = \dfrac{rK}{4}$. In Section 5.3.2 we show that the equilibrium values for the constant effort model are given by

$$P^* = \frac{K \pm \sqrt{K^2 - 4\frac{Kh}{r}}}{2}.$$

We plug in $h = \dfrac{rK}{4}$ and simplify:

$$P^* = \frac{K \pm \sqrt{K^2 - 4\frac{K}{r} \cdot \frac{rK}{4}}}{2}$$

$$P^* = \frac{K \pm \sqrt{K^2 - K^2}}{2}$$

$$P^* = \frac{K}{2}.$$

Thus when harvesting at the MSY in the constant take case, we end up with a single equilibrium value at half the carrying capacity.

5.5   THE RICKER MODEL

1 Use the Ricker model to project the salmon population from Example 5.17 over 20 years if the population starts at 50,000.

From Example 5.17 we have $r = 0.61$ and $K = 477,700$. Plugging these values into the Ricker Excel spreadsheet and dragging the formulas down to year $t = 20$ produces the results shown graphically in Figure 5.7. The horizontal line represents the carrying capacity, so we see that the salmon population is projected to be at carrying capacity by year 20.

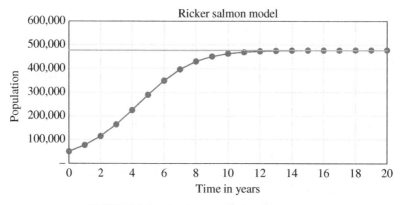

FIGURE 5.7   Excel output for Exercise 5.5.1.

3   *Extension*: Modify the Ricker model to include constant take harvesting using the MSY from Exercise 5.5.2. Investigate the effects of this harvesting level by projecting the fate of several different initial salmon populations.

The DDS for the Ricker model with constant take harvesting is given by

$$P(t) = (1+r)P(t-1)e^{-\frac{\ln(1+r)}{K}P(t-1)} - h.$$

Using the parameters from Example 5.17 and the $h = MSY = 64,216.7$ gives the model

$$P(t) = 1.61 \cdot P(t-1)e^{-\frac{\ln(1.61)}{477,700}P(t-1)} - 64,216.7.$$

We modify the Ricker Excel spreadsheet to account for the constant take harvesting. The resulting set-up with the formula for population displayed is given in Figure 5.8.

	A	B	C	D
1	Ricker Model with Constant Take Harvesting			
2				
3	Intrinsic growth rate, r =		0.61	
4	Carrying capacity, K =		477,700	
5	Constant take harvest, h =		64,217	
6				
7	t	Population		
8	0	200,000		
9	1	=(1+$C$3)*B8*EXP(-LN(1+$C$3)*B8/$C$4)-$C$5		
10	2	199,129.0		
11	3	198,655.7		
12	4	198,154.8		

FIGURE 5.8   Excel set-up for Exercise 5.5.3.

Numerical experiments with different initial populations reveal that if the initial population starts above about 240,000, then the population will decrease to about 240,000. If the initial population starts below about 240,000 but above about 210,000, then the population will increase back to 240,000. However, if the population starts below about 210,000 then the population will crash.

Just as it did for the discrete logistic model, the introduction of constant take harvesting near the MSY has resulted in two positive equilibrium values with the larger one being stable and the smaller one unstable. As we increase the harvesting level, the two equilibrium values eventually coalesce into a single semi-stable equilibrium. This happens at a harvesting level of approximately $h = 64,500$. Past that point the population will no longer have an equilibrium value and will crash regardless of the initial population.

5   *Extension*: Show that under a constant effort harvesting level of e, the Ricker model's positive equilibrium value becomes

$$P^* = K\left(1 - \frac{\ln(1+e)}{\ln(1+r)}\right).$$

Note the similarity with the constant effort harvesting equilibrium for the discrete logistic model.

We must solve for P^* such that

$$P^* = (1+r)P^* e^{-\frac{\ln(1+r)}{K}P^*} - eP^*.$$

We get

$$1 = (1+r)e^{-\frac{\ln(1+r)}{K}P^*} - e$$

$$\frac{(1+e)}{(1+r)} = e^{-\frac{\ln(1+r)}{K}P^*}.$$

Taking the natural logarithm of both sides yields

$$\ln\left[\frac{(1+e)}{(1+r)}\right] = -\frac{\ln(1+r)}{K}P^*.$$

Next we apply the logarithm rule that the logarithm of a quotient is equal to the difference of the logarithms:

$$\ln(1+e) - \ln(1+r) = -\frac{\ln(1+r)}{K}P^*.$$

Finally we get

$$-\frac{K}{\ln(1+r)}\left(\ln(1+e)-\ln(1+r)\right)=P^*$$

$$K\left(1-\frac{\ln(1+e)}{\ln(1+r)}\right)=P^*.$$

As an example we can check our result in Exercise 5.5.4. With $e=0.269$, $r=0.61$, and $K=477,700$, we have

$$P^*=K\left(1-\frac{\ln(1+e)}{\ln(1+r)}\right)=477,700\left(1-\frac{\ln(1.269)}{\ln(1.61)}\right)\approx 238,738.$$

This is very close to the estimate of 240,000 given in Exercise 5.5.4.

6

BLOOD ALCOHOL CONCENTRATION AND PHARMACOKINETICS

6.1 BLOOD ALCOHOL CONCENTRATION (BAC)

1 Scott is a 200-pound man who consumes 5 standard glasses of wine. Estimate Scott's BAC.

Here we use our basic Excel spreadsheet for calculating BAC. The set-up and result are shown in Figure 6.1. Scott's BAC would be approximately 0.11%.

	A	B	C
1	BAC Calculator		
2			
3	Body weight in lbs. =		200
4	Sex =		M
5	Number of drinks =		5
6			
7	BAC =	0.107	

FIGURE 6.1   Excel output for Exercise 6.1.1.

3 Anna is a 150-pound woman. Determine how many standard drinks she can consume and still remain below the legal driving limit for BAC.

Solutions Manual to Accompany Models for Life: An Introduction to Discrete Mathematical Modeling with Microsoft® Office Excel®, First Edition. Jeffrey T. Barton.
© 2016 John Wiley & Sons, Inc. Published 2016 by John Wiley & Sons, Inc.
Companion website: www.wiley.com/go/barton/solutionsmanual_modelsforlife

We use 0.08% as the legal limit for BAC, and we use Goal Seek with the BAC calculator spreadsheet to determine the corresponding number of drinks for Anna. The set-up just before running Goal Seek is shown in Figure 6.2. Figure 6.3 shows the result: Anna can have up to about 2.36 standard drinks while still remaining below the legal limit for driving.

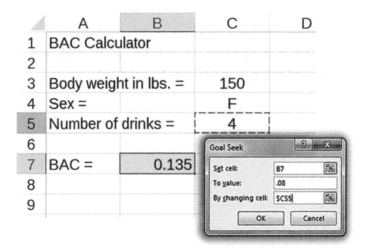

FIGURE 6.2   Excel Goal Seek set-up for Exercise 6.1.3.

	A	B	C
1	BAC Calculator		
2			
3	Body weight in lbs. =		150
4	Sex =		F
5	Number of drinks =		2.36
6			
7	BAC =	0.080	

FIGURE 6.3   Goal Seek result for Exercise 6.1.3.

5   *Extension*: Jeff is a 180-pound man who consumes two 1.5-oz. drinks of 100-proof Scotch. Estimate his BAC. You may use the fact that the mass density of alcohol is 23.3 g per fluid ounce.

Here we must estimate the amount of alcohol in the drinks Jeff has. An alcoholic beverage that is "100-proof" contains 50% alcohol, thus each 1.5-oz. drink of the Scotch contains $50\% \times 1.5 = 0.75$ fluid ounces of alcohol. Using the given mass density of alcohol, we calculate that there are $23.3 \times 0.75 = 17.475$ g of alcohol in each of Jeff's drinks. Note that this value is higher than the assumed amount, 14 g, in a standard drink.

Next we modify the BAC calculator spreadsheet so that it uses 17.475 g of alcohol per drink rather than 14 g. This is a matter of editing the formula for BAC in the

	A	B	C	D	E	F	G
1	BAC Calculator						
2							
3	Body weight in lbs. =		180				
4	Sex =		M				
5	Number of drinks =		2.00				
6							
7	BAC =		=((C5*17.475)/((C3/2.2046)*IF(C4="M",0.58,0.49)))*0.806/10				
8							

FIGURE 6.4   Excel set-up for Exercise 6.1.5.

	A	B	C
1	BAC Calculator		
2			
3	Body weight in lbs. =		180
4	Sex =		M
5	Number of drinks =		2.00
6			
7	BAC =	0.059	

FIGURE 6.5   Excel output for Exercise 6.1.5.

spreadsheet. The modification is shown in Figure 6.4. The result is shown in Figure 6.5: Jeff's BAC will be about 0.06% after 2 drinks of 100-proof Scotch.

7 Calculate your own BAC after having two drinks of your choosing.

If the chosen drinks are standard drinks, then we can just enter the appropriate body weight and sex into the BAC calculator spreadsheet. If the chosen drinks are not standard drinks, then the method described in Exercise 6.1.5 may be used.

9 *Extension*: Using the body water percentage estimate from Exercise 6.1.8, estimate your BAC after having the two drinks you used in Exercise 6.1.7. How much difference did the body water percentage estimate make for the BAC calculation?

For example purposes, we assume that Scott from Exercise 6.1.1 is a 30-year old black man who is 5' 10" tall. As Figure 6.6 shows, Scott's estimated BAC is

	A	B	C	D	E
1	BAC Calculator with Improved Body Water Estimate				
2					
3	Body weight in lbs. =		200		
4	Height in inches =		70		
5	Age in years =		30		
6	Sex =		M		
7	Race =		B		
8	Number of drinks =		2		
9					
10	BAC =	0.042			

FIGURE 6.6   Excel output for Exercise 6.1.9.

now 0.042% based on the Chumlea formula for TBW. Based on the original BAC calculator, Scott's predicted BAC would have instead been 0.043%. Because the new estimate is slightly lower than the original, the Chumlea formula must predict slightly higher TBW for Scott than the original estimate of 58% of his body weight.

6.2 THE WIDMARK MODEL

1 Give an explicit formula for the Widmark model.

With time in minutes the DDS for the Widmark model is given by

$$BAC(t) = BAC(t-1) - 0.000283.$$

Thus after 1 min we have $BAC(1) = BAC(0) - 0.000283$. After 2 min we get

$$BAC(2) = BAC(1) - 0.000283$$
$$BAC(2) = (BAC(0) - 0.000283) - 0.000283$$
$$BAC(2) = BAC(0) - 0.000283 \cdot 2.$$

Similarly, after 3 min we get

$$BAC(3) = BAC(2) - 0.000283$$
$$BAC(3) = (BAC(0) - 0.000283 \cdot 2) - 0.000283$$
$$BAC(3) = BAC(0) - 0.000283 \cdot 3.$$

After 4 min the result would be

$$BAC(4) = BAC(0) - 0.000283 \cdot 4.$$

Recognizing the general pattern, we have the explicit formula for the Widmark model:

$$BAC(t) = BAC(0) - 0.000283 \cdot t.$$

3 Use Excel's Data Validation feature to have the Widmark Excel model use different elimination rates based on user-selected drinking habits. In particular, use a drop-down menu to allow the user to select light, moderate, or heavy drinker and have the BAC calculation use the corresponding elimination rate.

According to the text, the BAC elimination rate is typically lower for light drinkers and higher for heavy drinkers, with a standard range being a decrease of 0.10 to 0.20

BAC per hour. We use 0.10 for the elimination rate of light drinkers, 0.017 for moderate drinkers, and 0.20 for heavy drinkers. Since our time units are minutes for the Widmark model, we divide the corresponding rate by 60 in our Excel formula.

The Excel set-up with the drop-down menu for drinking habits displayed is given in Figure 6.7.

	A	B	C
1	Widmark BAC Model		
2			
3	Body weight in lbs. =		180
4	Gender =		M
5	Drinking behavior =		
6	Number of drink	Light	
7		Moderate	
		Heavy	

FIGURE 6.7 Excel Data Validation set-up for Exercise 6.2.3.

Finally we incorporate the different rates into our Excel formula by using a nested IF statement. The new formula is shown in Figure 6.8.

	A	B	C	D	E	F	G
1	Widmark BAC Model						
2							
3	Body weight in lbs. =		180				
4	Gender =		M				
5	Drinking behavior =		Moderate				
6	Number of drinks =		2				
7							
8		t	BAC(t)				
9		0	0.048				
10		1	=B9-IF($C$5="Light",0.01,IF($C$5="Moderate",0.017,0.02))/60				
11		2	0.047				

FIGURE 6.8 Excel set-up for Exercise 6.2.3.

5 Use your own parameters and drink of choice to:

a. Estimate your initial BAC.

For example purposes, suppose the user is a 140-pound man who has 3 standard drinks. Plugging these parameters into the Widmark model produces the result shown in Figure 6.9. The initial BAC would be 0.092%.

b. Estimate how long it will take before you are legally able to drive.

Taking the legal driving limit to be 0.08%, we copy the Excel formula down until the above BAC falls below that level. The result with most rows hidden is

	A	B	C
1	Widmark BAC Model		
2			
3	Body weight in lbs. =		140
4	Sex =		M
5	Number of drinks =		3
6			
7	*t*	*BAC(t)*	
8	0	0.092	
9	1	0.092	
10	2	0.091	

FIGURE 6.9   Excel output for Exercise 6.2.5a.

	A	B	C
1	Widmark BAC Model		
2			
3	Body weight in lbs. =		140
4	Sex =		M
5	Number of drinks =		3
6			
7	*t*	*BAC(t)*	
8	0	0.092	
9	1	0.092	
51	43	0.080	
52	44	0.079	
53	45	0.079	

FIGURE 6.10   Excel output for Exercise 6.2.5b.

presented in Figure 6.10. Assuming no more alcohol is consumed, the Widmark model projects that it will take about 45 min before the person in part a. is legally able to drive.

7 *Extension*: Graph the Widmark model BAC projections over time for parameters of your choosing. On the Excel graph, include horizontal lines marking the BAC levels for the legal driving limit, drunkenness, coma, and death.

Referring to Text Table 6.1, we use 0.08% for the legal driving limit, 0.15% for drunkenness, 0.30% for coma, and 0.40% for death. To get horizontal lines to appear on an Excel graph, we create columns for each level where every entry in the column is the same value. The set-up is shown in Figure 6.11.

To create the graph, we select all columns including the headings and select a scatter plot. Figure 6.12 shows the resulting graph for a 140-pound male who consumes 6 standard drinks.

TEXT TABLE 6.1   The Effects of Alcohol at Given BAC

BAC	Effects
0.02–0.03	Mild relaxation, mild euphoria, some loss of judgment, decline in ability to multitask
0.04–0.06	Lowered inhibition, feeling of euphoria, emotions intensified, impaired judgment, impaired coordination, loss of reflexes
0.08 (legal limit for driving)	Impaired balance, impaired judgment, loss of self-control, loss of concentration, poor speed control and perception while driving
0.10–0.125	Poor reaction time and control, slurred speech, poor coordination, inability to maintain lane position while driving
0.13–0.15	Further loss of muscle control and balance, possible vomiting, substantial impairment of ability to control vehicle
0.16–0.19	Possible vomiting, drinker appears very drunk
0.20	Vomiting, blackouts, inability to stand or walk without assistance, increased likelihood of injury, risk of choking on vomit
0.30–0.35	Stupor, passing out, coma possible, death possible
0.40	Coma, death likely

	A	B	C	D	E	F
1	Widmark BAC Model					
2						
3	Body weight in lbs. =		140			
4	Sex =		M			
5	Number of drinks =		3			
6						
7	t	BAC(t)	Driving	Drunk	Coma	Death
8	0	0.092	0.08	0.15	0.3	0.4
9	1	0.092	0.08	0.15	0.3	0.4
10	2	0.091	0.08	0.15	0.3	0.4
11	3	0.091	0.08	0.15	0.3	0.4

FIGURE 6.11   Excel set-up for Exercise 6.2.7.

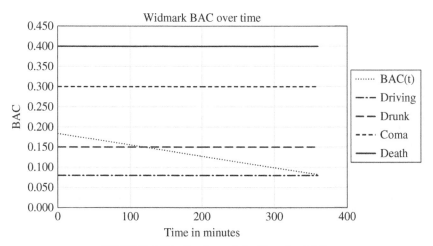

FIGURE 6.12   Excel graph for Exercise 6.2.7.

6.3 THE WAGNER MODEL

1 Implement the Wagner model in Excel. Allow the user to input sex, number of drinks consumed, and body weight as parameters in the model. Store V_{max} and K_m in their own cells for easy experimentation later.

The Excel set-up with the Wagner formula displayed is given in Figure 6.13.

	A	B	C	D
2				
3	Body weight in lbs. =		150	
4	Sex =		F	
5	Number of drinks =		3	
6	V_{max} =		0.000369	
7	K_m =		0.00858	
8				
9	t	BAC(t)		
10	0	0.102		
11	1	=B10-($C$6/($C$7+B10))*B10		
12	2	0.101		
13	3	0.101		

FIGURE 6.13   Excel set-up for Exercise 6.3.1.

3 Use the Wagner model with your own parameter choices to:
   a. Graph a projection of your BAC over time.

For example purposes we use a 200-pound male who consumes 4 standard drinks. After entering these values into the Wagner Excel model, we obtain the graph shown in Figure 6.14.

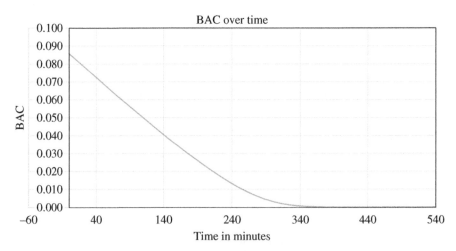

FIGURE 6.14   Excel graph for Exercise 6.3.3.

b. Estimate how long it will take before you are legally able to drive.

Here we can look at the graph or the Excel output to estimate when our BAC will fall below the usual legal driving limit of 0.08%. According to the graph, this appears to happen at approximately minute 20. The model projects that it will take about 20 min for us to legally be able to drive.

5 *Extension*: Make the Wagner Excel model more flexible by allowing users to input non-standard drinks.

There are many ways to approach this problem in Excel, but all of them will involve having the user or Excel calculate the total amount of alcohol in grams that the user has consumed. Since most alcoholic beverages report alcohol content as a percentage by volume, the weight density of alcohol is necessary: 23.3 g per fluid ounce. As an example of the sort of calculation that is necessary, suppose someone consumes 4 ounces of 110-proof whiskey. Since 110-proof whiskey is 55% alcohol, this would amount to consuming $4 \times 55\% = 2.2$ fluid ounces of alcohol. Thus a total of $2.2 \times 23.3 = 51.26$ g of alcohol would be consumed. Figure 6.15 shows how to automate the calculation in Excel.

	A	B	C	D	E	F	G
1	Wagner BAC Model						
2							
3	Body weight in lbs. =		200	Ounces of beverage consumed =			4
4	Sex =		M	% Alcohol by volume =			55%
5	V_{max} =		0.000369	Total grams alcohol consumed =			=G3*G4*23.3
6	K_m =		0.00858				

FIGURE 6.15 Excel total alcohol calculation set-up for Exercise 6.3.5.

Once the total grams of alcohol from all drinks is known, it can be substituted in the initial BAC calculation instead of the previous expression "number of standard drinks times 14 g." The modified formula is displayed in Figure 6.16.

	A	B	C	D	E	F	G
1	Wagner BAC Model						
2							
3	Body weight in lbs. =		200	Ounces of beverage consumed =			4
4	Sex =		M	% Alcohol by volume =			55%
5	V_{max} =		0.000369	Total grams alcohol consumed =			51.26
6	K_m =		0.00858				
7							
8	t	BAC(t)					
9		0	=((G5)/((C3/2.2046)*IF(C4="M",0.58,0.49)))*0.806/10				
10	1	0.078					

FIGURE 6.16 Excel set-up for Exercise 6.3.5.

7   In a DUI court case, a defendant's BAC is measured at the scene of an accident as
    0.05. The defendant is a 120-pound woman who is known to have stopped drink-
    ing 3 h before being tested. At the time of the accident she had been driving for
    30 min. Use the Wagner model to determine whether she is guilty of DUI.

    We need to know the woman's BAC during the 30 min she was driving: from
    $t = 150$ to $t = 180$. In order to use the Wagner model to estimate this, we must deter-
    mine how many drinks the woman had. This is a Goal Seek problem where we

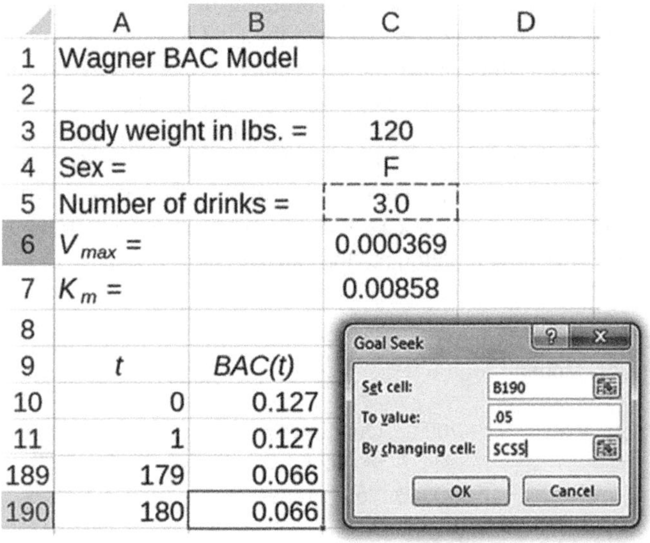

FIGURE 6.17   Excel Goal Seek set-up for Exercise 6.3.7.

FIGURE 6.18   Goal Seek result for Exercise 6.3.7.

need to find the number of drinks that would lead to a BAC of 0.05% at $t = 180$. The Excel set-up before Goal Seek is run is shown in Figure 6.17. Figure 6.18 shows the result: the Wagner model estimates that the defendant had about 2.6 standard drinks.

With the number of drinks set to 2.6, we examine the defendant's projected BAC level at the time she started driving, i.e. at $t = 150$. As Figure 6.19 shows, the defendant's BAC level was 0.06% at the time she began driving, so using the standard legal limit of 0.08%, she is not guilty of DUI.

	A	B	C
1	Wagner BAC Model		
2			
3	Body weight in lbs. =		120
4	Sex =		F
5	Number of drinks =		2.6
6	V_{max} =		0.000369
7	K_m =		0.00858
8			
9	t	$BAC(t)$	
10	0	0.110	
11	1	0.109	
159	149	0.060	
160	150	0.060	

FIGURE 6.19 Excel output for Exercise 6.3.7.

9 Compare the average maximal elimination rate for the Wagner model with the average elimination rate for the Widmark model.

The average maximal elimination rate used in the Wagner model is $V_{max} = 0.0003693$. The units for V_{max} are BAC per minute. To compare this with the average Widmark elimination rate, we multiply by 60 to get a rate of $0.0003693 \cdot 60 \approx 0.022$ BAC per hour. The average elimination rate used in the Widmark model is 0.017 BAC per hour, so the maximal rate in the Wagner model is higher. In Figure 6.20 we verify this by showing both model projections on the same graph for a 180-pound female who has 5 drinks. The Wagner model projection is below the Widmark projection for most of the time.

It makes sense that the Widmark rate is lower than the maximal Wagner rate. The Wagner model assumes a varying elimination rate that is higher at higher BAC levels and lower at lower BAC levels. Since the Widmark model must use a single average rate, it should not be surprising that it is below Wagner's maximum rate.

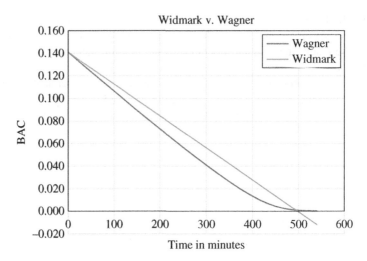

FIGURE 6.20   Excel graph for Exercise 6.3.9.

6.4   ALCOHOL CONSUMPTION PATTERNS

1   Toni is a 170 pound woman who drinks 3 standard drinks per hour. Project Toni's BAC at the end of 4 h using the modified Wagner model.

We plug all parameters into the Excel spreadsheet for the Wagner model with constant drinking rate, and we copy the formulas down to $t = 240$. Figure 6.21 shows the projected BAC for Toni with most rows hidden. At the end of 4 h the model projects that Toni's BAC will be about 0.28%, a level that puts her at risk of serious health consequences including coma and death.

	A	B	C	D
1	Wagner BAC Model with Constant Drinking Rate			
2				
3	Body weight in lbs. =		170	
4	Gender =		F	
5	Drinks per hour =		3	
6	BAC incr. min^{-1}, d =		0.0014932	
7	V_{max} =		0.0003693	
8	K_m =		0.00858	
9				
10	t	BAC(t)		
11	0	0.000		
12	1	0.001		
250	239	0.278		
251	240	0.279		

FIGURE 6.21   Excel output for Exercise 6.4.1.

3   Using average values for the Wagner parameters, estimate how many drinks per hour you would need to have in order to maintain a BAC level of 0.045.

For example purposes we suppose that the user is a 135-pound female. The first step is to use the formula for BAC^* to find d as we did in Example 6.9. If we want to maintain a BAC level of 0.045, we need $BAC^* = 0.045$. Thus we need to calculate

$$d = \frac{V_{\max} \cdot BAC^*}{K_m + BAC^*} = \frac{0.0003693 \cdot 0.045}{0.00858 + 0.045} \approx 0.0003102.$$

The question now is how many drinks per hour our example woman needs to consume in order for her BAC to increase by $d \approx 0.0003102$ per min. We find this using Goal Seek and the modified Wagner spreadsheet. The Excel set-up just before running goal Seek is given in Figure 6.22. The result of a successful Goal Seek is shown in Figure 6.23. Our example woman would need to consume about 0.495 drinks per h to maintain a BAC level of 0.045.

	A	B	C	D
1	Wagner BAC Model with Constant Drinking Rate			
2				
3	Body weight in lbs. =		135	
4	Gender =		F	
5	Drinks per hour =		1	
6	BAC incr. min^{-1}, d =		0.0006268	
7	V_{max} =		0.0003693	
8	K_m =		0.00858	
9				
10	t	$BAC(t)$		
11	0	0.000		
12	1	0.001		

Goal Seek

Set cell: C6
To value: .0003102
By changing cell: $C$5

OK   Cancel

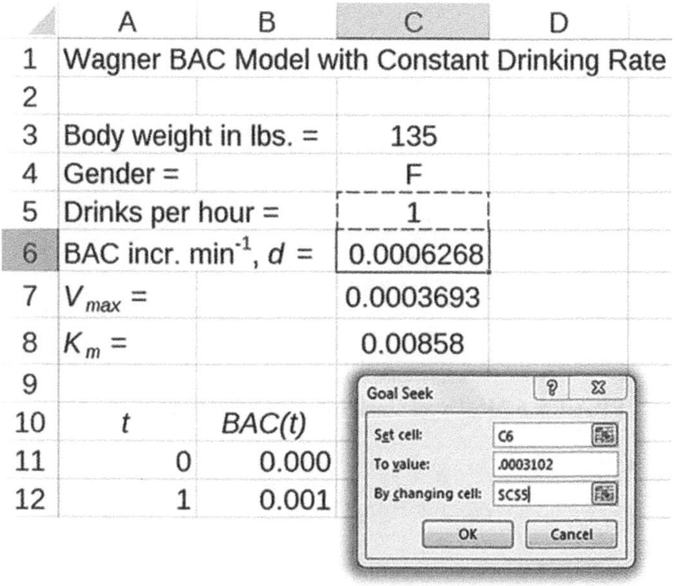

FIGURE 6.22   Excel Goal Seek set-up for Exercise 6.4.3.

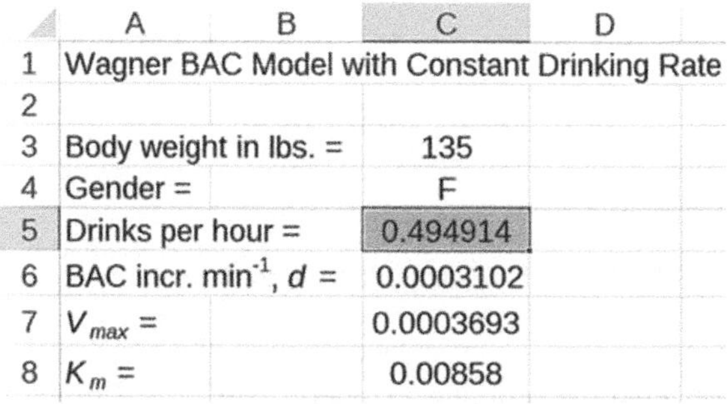

	A	B	C	D
1	Wagner BAC Model with Constant Drinking Rate			
2				
3	Body weight in lbs. =		135	
4	Gender =		F	
5	Drinks per hour =		0.494914	
6	BAC incr. min^{-1}, d =		0.0003102	
7	V_{max} =		0.0003693	
8	K_m =		0.00858	

FIGURE 6.23   Goal Seek result for Exercise 6.4.3.

5   What would the loading dose need to be in order for you to quickly raise your BAC to 0.045?

To find the loading dose for the woman in Exercise 6.4.3 we need to use the original Wagner BAC spreadsheet and Goal Seek to find the number of drinks corresponding to an initial BAC of 0.045%. The Excel set-up before running Goal Seek is shown in Figure 6.24. Figure 6.25 shows the result of a successful Goal Seek. The woman would need to consume about 1.2 drinks initially to quickly raise her BAC up to 0.045%.

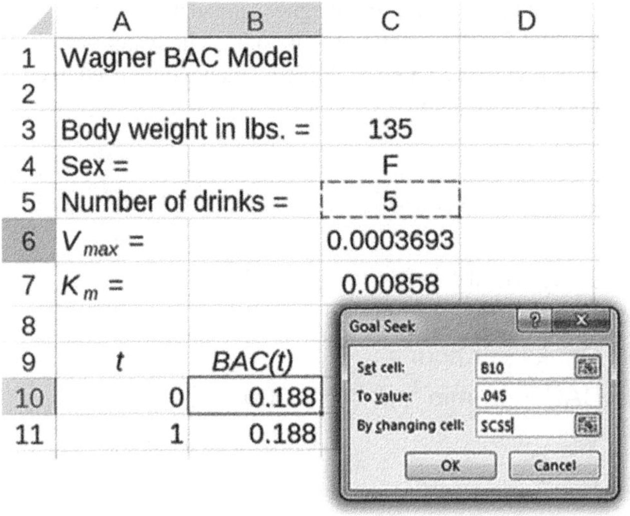

FIGURE 6.24   Excel Goal Seek set-up for Exercise 6.4.5.

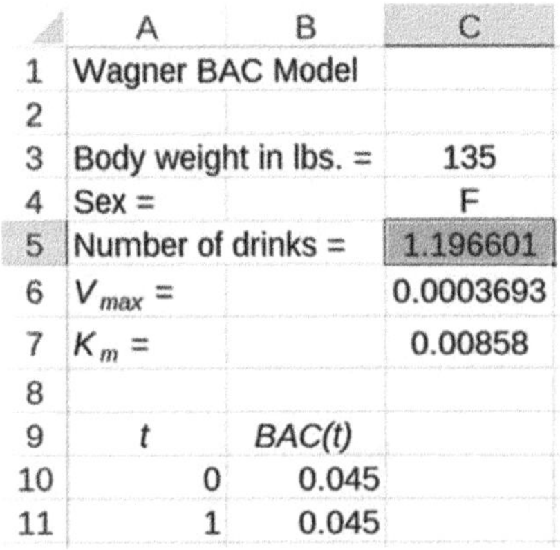

FIGURE 6.25   Goal Seek result for Exercise 6.4.5.

7   In Exercise 6.4.3 suppose you accidentally drink 20% more per hour than your calculated rate. Find the resulting long-term BAC. Explain the significance of this result.

Using the results from Exercise 6.4.3, drinking 20% more than intended means consuming $0.495 + 20\% \times 0.495 = 0.594$ drinks per h rather than 0.495 drinks per h. This change would result in an increased value for d: $d \approx 0.0003723$. The increase in d would in turn bring about an increase in the long-term BAC:

$$BAC^* = \frac{dK_m}{V_{max} - d} = \frac{0.0003723 \cdot 0.00858}{0.0003693 - 0.0003723} \approx -1.065.$$

This value is not physically meaningful, and it is an example of what happens when the drinking rate exceeds the body's ability to eliminate the incoming alcohol. The woman in this example would see her BAC continue to rise without bound until negative health effects prevented her from drinking further. Thus increasing her drinking rate even by a modest amount would change her long-term fate from being buzzed to experiencing dangerous health effects.

9 *Extension*: In the same manner as we did for the Wagner model, incorporate a drinking rate and drinking time into the Widmark Excel model. Compare this modified Widmark model with the modified Wagner model for parameter values of your choosing.

Here we can take the Wagner model with drinking time spreadsheet and modify it so that the BAC column uses the Widmark formula instead of the Wagner formula. Figure 6.26 shows the Excel set-up with the required Widmark formula showing. To compare the two models we graph them on the same axes. Figure 6.27 shows the graph over a period of 9 h for a 150-pound female who consumes 4 drinks over a period of 2 h. The graphs are virtually identical as BAC increases, but once drinking stops we see that the Wagner graph decreases faster than the Widmark graph for most of the remainder of the time. The graphs appear to reach the BAC level of 0 at roughly the same time.

	A	B	C	D
1	Widmark BAC Model with Drinking Time			
2				
3	Body weight in lbs. =		205	
4	Gender =		M	
5	Drinks consumed =		6	
6	Drinking time, DT =		300	
7	BAC incr. min^{-1}, d =		0.0004184	
8				
9				
10				
11		t	$BAC(t)$	
12		0	0.000	
13		1	=B12-0.000283+IF(A13<$C$6,$C$7,0)	

FIGURE 6.26 Excel set-up for Exercise 6.4.9.

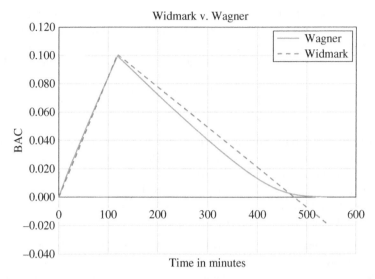

FIGURE 6.27   Excel graph for Exercise 6.4.9.

6.5   MORE GENERAL DRUG ELIMINATION

1 Create an Excel spreadsheet that will automatically calculate the elimination rate, r, from the percentage of drug eliminated over a specified period of time. The user should be able to enter both the percentage eliminated and the time period over which it happens.

With r as the elimination rate, recall that the amount of drug left in the body at time t is given by

$$B(t) = (1-r)^t B(0).$$

If we let ρ be the percentage eliminated in T_ρ minutes, then with ρ as a decimal we know

$$B(T_\rho) = (1-r)^{T_\rho} B(0)$$
$$(1-\rho)B(0) = (1-r)^{T_\rho} B(0).$$

Solving for r gives

$$(1-\rho) = (1-r)^{T_\rho}$$

$$(1-\rho)^{\frac{1}{T_\rho}} = 1-r$$

$$r = 1-(1-\rho)^{\frac{1}{T_\rho}}.$$

	A	B	C	D
1	Finding the Elimination Rate			
2				
3	Percent eliminated, ρ =		50%	
4	Time period, T_ρ =		120	(minutes)
5	Elimination rate, r =		=1-(1-C3)^(1/C4)	

FIGURE 6.28 Excel set-up for Exercise 6.5.1.

What remains is to type this formula into Excel where ρ and T_ρ are parameters input by the user. The Excel set-up with the formula for r displayed is given in Figure 6.28.

3 Suppose you know that 95% of a drug is eliminated in a 24-h period. Determine the elimination rate, r, for the drug.

We can find this algebraically as in the text or we can use Excel, specifically the spreadsheet create in Exercise 6.5.1. Using the spreadsheet from Exercise 6.5.1 we get the set-up shown in Figure 6.29. The time units here are hours. The elimination rate for the drug is thus $r \approx 11.7\%$ per h.

	A	B	C
1	Finding the Elimination Rate		
2			
3	Percent eliminated, ρ =		95%
4	Time period, T_ρ =		24
5	Elimination rate, r =		0.1173

FIGURE 6.29 Excel set-up for Exercise 6.5.3.

5 The antidepressant Zoloft has an elimination rate of $r = 2.63\%$ per h, and it reaches peak plasma concentration roughly 6 h after ingestion (RxList, 2015). Determine the absorption rate, α.

We use our two-compartment drug model to produce a graph of $B(t)$, and then we experiment with different values for α until the peak of the $B(t)$ graph is above time $t = 6$. Figure 6.30 shows the Excel set-up along with the appropriate graph. With α set at 40% per h we get the peak of $B(t)$ occurring at the appropriate time, $t = 6$ h.

7 Suppose a drug with typical dissolution time is prescribed in pill form so that the patient takes a 300-mg pill once every 6 h. Find the dosing function.

With minutes as our time units and frequency, f, in hours, the general dosing function has the form

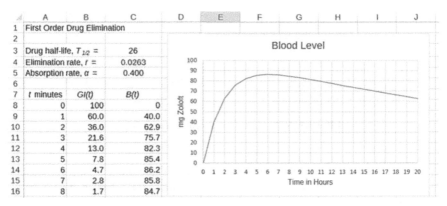

	A	B	C	D	E	F	G	H	I	J
1	First Order Drug Elimination									
2										
3	Drug half-life, $T_{1/2}$ =		26							
4	Elimination rate, r =		0.0263							
5	Absorption rate, α =		0.400							
6										
7	t minutes	$GI(t)$	$B(t)$							
8	0	100	0							
9	1	60.0	40.0							
10	2	36.0	62.9							
11	3	21.6	75.7							
12	4	13.0	82.3							
13	5	7.8	85.4							
14	6	4.7	86.2							
15	7	2.8	85.8							
16	8	1.7	84.7							

FIGURE 6.30   Excel set-up for Exercise 6.5.5.

$$D(t) = \begin{cases} d/\delta & \text{if } 0 \le t \mod(60f) < \delta \\ 0 & \text{otherwise} \end{cases}.$$

In the formula for the dosing function, d represents the dosage and δ the dissolution time. A typical dissolution time is 30 min, and with the other parameters plugged in, the dosing function becomes

$$D(t) = \begin{cases} 10 & \text{if } 0 \le t \mod(360) < 30 \\ 0 & \text{otherwise} \end{cases}.$$

9   *Extension*: Suppose a patient takes one 100 mg pill of Zoloft every day. If each pill takes 30 min to dissolve, project the amount of Zoloft in the body over 1 week. Use hours as the time unit.

Here we use the parameters for Zoloft given in the previous exercises. Since those parameters assumed hours for the time unit we have to modify our spreadsheet, in particular the dosing function. If we use hours as the time unit, the dissolution time will be less than one unit. Thus as a best approximation we use $\delta = 1$ h. In the formula for the dosing function, we also need to eliminate the multiplication of f by 60 since we are no longer converting the frequency to minutes. The Excel set-up with the dosing function formula displayed is given in Figure 6.31. The time of interest is 1 week after the first dose, so $t = 7 \times 24 = 168$ h. The result is shown in Figure 6.32. There will be about 118 mg of Zoloft in the body.

We note that we could get a slightly better projection by using half-hours as our time unit. This has the benefit of making a dissolution time of 1 accurate, but it has the disadvantage of making us re-calculate the elimination and absorption rates. The reader can check that the difference in the final answer is negligible. Using half-hour as the time unit produces a projection of 117.4 mg of Zoloft a week after the first dose, a difference of less than 1 mg from the previous estimate.

	A	B	C	D	E	F	G
1	Two-Compartment Drug Model with Dosing Function and Dissolution Time						
2							
3	Drug half-life, $T_{1/2}$ =		26	(hours)			
4	Elimination rate, r =		0.026307				
5	Absorption rate, α =		0.400				
6	Dosage amount, d =		100.0	(mg)			
7	Dose frequency, f =		24.0	(hours)			
8	Dissolution time, δ =		1.0	(hours)			
9							
10	t minutes	$D(t)$	$GI(t)$	$B(t)$			
11		0	=IF(MOD(A11,$C$7)<$C$8,$C$6/$C$8,0)				

FIGURE 6.31   Excel set-up for Exercise 6.5.9.

	A	B	C	D	E	F
1	Two-Compartment Drug Model with Dosing Function and Dissolution Time					
2						
3	Drug half-life, $T_{1/2}$ =		26	(hours)		
4	Elimination rate, r =		0.026307			
5	Absorption rate, α =		0.400			
6	Dosage amount, d =		100.0	(mg)		
7	Dose frequency, f =		24.0	(hours)		
8	Dissolution time, δ =		1.0	(hours)		
9						
10	t minutes	$D(t)$	$GI(t)$	$B(t)$		
11	0	100	100.0	0		
12	1	0	60.0	40.0		
178	167	0	0.0	121.3		
179	168	100	100.0	118.1		

FIGURE 6.32   Excel output for Exercise 6.5.9.

6.6   THE VOLUME OF DISTRIBUTION

1   The antidepressant Zoloft (sertraline) is widely distributed throughout the body tissues and as a result has a large volume of distribution: 20 l per kg. Determine the proportion of Zoloft that is found in the plasma.

From the text we know that the proportion of a drug found in the plasma is given by

$$\text{proportion in plasma} = \frac{\text{actual plasma volume}}{\text{volume of distribution}}.$$

The estimate used in the text for actual plasma volume in humans is 0.04 l per kg of body weight. Thus we get

$$\text{proportion Zoloft in plasma} = \frac{\text{actual plasma volume}}{\text{volume of distribution}} = \frac{0.04 \text{ l per kg}}{20 \text{ l per kg}} = 0.002.$$

3 Suppose we know 25% of a particular drug ends up in the plasma. Determine this drug's volume of distribution.

We have

$$\text{proportion in plasma} = \frac{\text{actual plasma volume}}{\text{volume of distribution}}$$

$$0.25 = \frac{0.04}{V_D}$$

$$V_D = \frac{0.04}{0.25} = 0.16 \text{ l per kg}.$$

5 A patient weighing 175 pounds has a plasma concentration of 10 mg per l. Determine how much drug in total is in the patient's body if the volume of distribution for the drug is 0.16 l per kg.

First we find the proportion of the drug that is in the plasma:

$$\text{proportion in plasma} = \frac{\text{actual plasma volume}}{\text{volume of distribution}} = \frac{0.04 \text{ l per kg}}{0.16 \text{ l per kg}} = 0.25.$$

Next we find the total amount of drug in the plasma. For this we need to estimate how much plasma the patient has. The patient weighs about 79.4 kg and hence has about

$$0.04 \text{ l per kg} \times 79.4 \text{ kg} = 3.176 \text{ l of plasma.}$$

At a concentration of 10 mg per l, the patient has a total of 10 mg per l × 3.176 l = 31.76 mg of the drug in the plasma. Since this 31.76 mg represents only 25% of the total amount of drug in the body, the total amount is given by 31.76 × 4 = 127.04 mg.

6.7 COMMON DRUGS

1 Jessie is sore from a long run and is going to take some ibuprofen. If Jessie weighs 120 pounds and takes 2, 200 mg tablets every 4 h, graph her ibuprofen plasma concentration along with horizontal lines for the therapeutic window.

	A	B	C	D	E	F	G	H
1	Two-Compartment Drug Model for Plasma Concentration							
2								
3	Drug half-life, $T_{1/2}$ =	120	(minutes)		Volume of dist., V_d =	0.12		(l/kg)
4	Elimination rate, r =	0.005760			Prop. in plasma, ρ =	0.33		
5	Absorption rate, α =	0.036			Body weight =	120		(pounds)
6	Dosage amount, d =	400.0	(mg)		Plasma volume =	2.2		(liters)
7	Dose frequency, f =	4.0	(hours)					
8	Dissolution time, δ =	30.0	(minutes)					
9	Total doses, N =	6						
10								
11	t minutes	$D(t)$	$GI(t)$	$B(t)$	$PC(t)$ mg/l	10 mg/l	50 mg/l	200 mg/l
12	0	13.33333	13.3	0	0	10	50	200
13	1	13.33333	26.2	0.5	0.07	10	50	200

FIGURE 6.33   Excel set-up for Exercise 6.7.1.

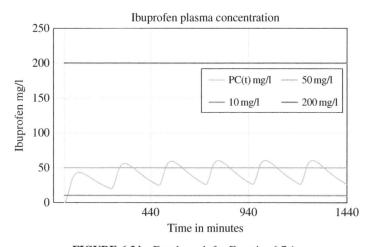

FIGURE 6.34   Excel graph for Exercise 6.7.1.

We use the Excel spreadsheet that was developed in the text for Example 6.24. Our dosage is now $d = 400$ and our frequency is $f = 4$. Figure 6.33 shows the Excel set-up with the relevant parameters entered. The graph is shown in Figure 6.34. Jessie's ibuprofen plasma concentration is for the most part in the center of the minimum recommended range for therapeutic effects.

3   Suppose Steve is a 210-pound Olympic athlete who consumes four strong 8-oz. cups of coffee consecutively, finishing the last 30 min before his competition. Will Steve's caffeine plasma concentration exceed the IOC limit? Explain.

We use the spreadsheet and the parameter values for caffeine given in Example 6.25. After plugging in the relevant parameters we get the set-up shown in Figure 6.35. The graph of Steve's caffeine plasma concentration is shown in

	A	B	C	D	E	F	G	H
1	Two-Compartment Drug Model for Plasma Concentration							
2								
3	Drug half-life, $T_{1/2}$ =		180	(minutes)	Volume of dist., V_d =	0.5		(l/kg)
4	Elimination rate, r =		0.003843		Prop. in plasma, ρ =	0.08		
5	Absorption rate, a =		0.0973		Body weight =	210		(pounds)
6	Dosage amount, d =		150.0	(mg)	Plasma volume =	3.8		(liters)
7	Dose frequency, f =		0.25	(hours)				
8	Dissolution time, δ =		15.0	(minutes)				
9	Total doses, N =		4					
10								
11	t minutes	$D(t)$		$GI(t)$	$B(t)$	$PC(t)$ mg/l	IOC max	
12	0	10		10.0	0	0	16.8	
13	1	10		19.0	1.0	0.02	16.8	
14	2	10		27.2	2.8	0.06	16.8	

FIGURE 6.35   Excel set-up for Exercise 6.7.3.

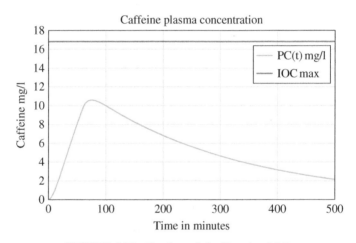

FIGURE 6.36   Excel graph for Exercise 6.7.3.

Figure 6.36. We note that Steve's caffeine plasma concentration level remains well below the IOC limit.

5   We based our elimination rate for caffeine on a value for the half-life that was near the middle of the given half-life range. The range was given as 2.5–4.5 h, and we used 3 h to find r. Investigate how much difference it would make in Example 6.25 if the half-life were a) 2.5 h, and b) 4.5 h.

The Excel set-up with all relevant parameters from Example 6.25 is shown in Figure 6.37. In that example, Mia's peak plasma concentration is about 15.45 mg per l. Changing the half-life to 150 min (2.5 h) yields a peak plasma concentration of about 15.09 mg per l. Changing the half-life to

	A	B	C	D	E	F	G	H
1	Two-Compartment Drug Model for Plasma Concentration							
2								
3	Drug half-life, $T_{1/2}$ =	180		(minutes)	Volume of dist., V_d =		0.5	(l/kg)
4	Elimination rate, r =	0.003843			Prop. in plasma, p =		0.08	
5	Absorption rate, α =	0.0973			Body weight =		110	(pounds)
6	Dosage amount, d =	150.0		(mg)	Plasma volume =		2.0	(liters)
7	Dose frequency, f =	0.25		(hours)				
8	Dissolution time, δ =	15.0		(minutes)				
9	Total doses, N =	3.0						

FIGURE 6.37   Excel set-up for Exercises 6.7.5 and 6.7.6.

270 min (4.5 h) yields a peak plasma concentration of about 16.12 mg per l. Neither of the changes would result in a projection that Mia would exceed the IOC limit.

7   *Extension*: Suppose Mia from Example 6.25 is taking oral contraceptives. Rework Exercise 6.7.4 taking into account this new information.

As stated in the text, oral contraceptives can have the effect of doubling the half-life of caffeine. Thus we take the half-life to be 360 min and re-do the trial and error approach from Exercise 6.7.4. This time we get 3.1 lattés as Mia's limit, which is not very different from the original value of 3.3 found in Exercise 6.7.4.

9   Use our Excel model to estimate how many 200 mg tablets of ibuprofen a 200-pound man can ingest at once before his ibuprofen plasma concentration reaches toxic levels.

We set the number of doses to 1 and use trial-and-error in Excel to determine the dosage level that would result in a plasma concentration above 200 mg per l. Figure 6.38 shows the plasma concentration graph will reach the toxic level

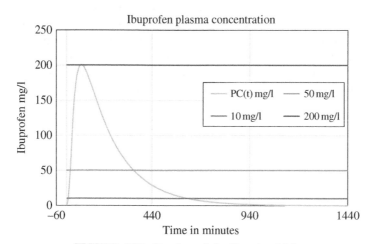

FIGURE 6.38   Excel graph for Exercise 6.7.9.

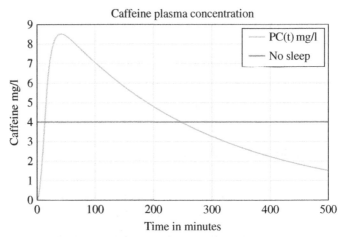

FIGURE 6.39   Excel graph for Exercise 6.7.11.

for a single dose of 3100 mg. This dosage level is equivalent to 15.5 tablets. We reiterate that while plausible, this result is based on many model assumptions that may not hold true in practice. It would not be a good idea to base any dosage decision on the results of this simple model. Always follow the directions provided with any medication.

11 *Extension:* Suppose a caffeine plasma concentration of 4 mg per l is enough to disturb your sleep. Estimate how late in the day you could have a grande Starbucks coffee and not disturb your sleep.

For example purposes we assume a body weight of 150 pounds, and we use 330 mg of caffeine for a Starbucks grande. We will also assume a bedtime of 11:00 p.m. For all other parameters we use the same values as in the text. The graph in Figure 6.39 shows our caffeine plasma concentration over time after consuming one Starbucks grande. Note that it takes about 250 min for the caffeine level to fall below the level for sleep disturbance. This is equivalent to 4 h and 10 min. Thus we would need to have our coffee no later than 6:50 p.m. in order to not have our sleep disturbed by the caffeine.

7

RANKING METHODS

7.1 INTRODUCTION TO MARKOV MODELS

1 Suppose a truck rental company has locations in Birmingham, Alabama, Columbia, South Carolina, and Dallas, Texas. The company permits one-way rentals and knows based on past experience that during a typical week the following truck movements occur.

Of the trucks that start in Birmingham:
- 50% stay in Birmingham;
- 25% travel to Columbia; and
- 25% travel to Dallas.

Of the trucks that start in Columbia:
- 40% stay in Columbia;
- 20% travel to Birmingham; and
- 40% travel to Dallas.

Of the trucks that start in Dallas:
- 80% stay in Dallas;
- 15% travel to Birmingham; and
- 5% travel to Columbia.

Solutions Manual to Accompany Models for Life: An Introduction to Discrete Mathematical Modeling with Microsoft® Office Excel®, First Edition. Jeffrey T. Barton.
© 2016 John Wiley & Sons, Inc. Published 2016 by John Wiley & Sons, Inc.
Companion website: www.wiley.com/go/barton/solutionsmanual_modelsforlife

a. Give a flow diagram for the model.

The flow diagram is shown in Figure 7.1.

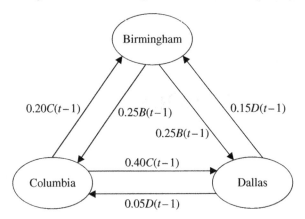

FIGURE 7.1 Flow diagram for Exercise 7.1.1.

b. Implement the model in Excel.

We use the Excel spreadsheet developed for Example 7.1, and we plug in the new percentages as the parameters. Figure 7.2 shows the Excel set-up.

	A	B	C	D	E
1	Truck Rental Markov Model				
2					
3	From Birmingham to Columbia =				25.0%
4	From Birmingham to Dallas =				25.0%
5	From Columbia to Birmingham =				20.0%
6	From Columbia to Dallas =				40.0%
7	From Dallas to Birmingham =				15.0%
8	From Dallas to Columbia =				5.0%
9					
10	t	Birmingham	Columbia	Dallas	
11	0	200	200	200	
12	1	170.0	140.0	290.0	
13	2	156.5	113.0	330.5	

FIGURE 7.2 Excel set-up for Exercise 7.1.1.

c. If 200 trucks begin in each city, use Excel to determine the long-term numbers of trucks in each city.

To find the long-term number of trucks in each city with Excel, we copy the formulas down until the numbers stop changing. For this example, it takes about 15 weeks for the numbers to stop changing, and we show the results in Figure 7.3. Rounding to the nearest truck, we see that we will end up with 145 trucks in Birmingham, 91 in Columbia, and 364 in Dallas.

	A	B	C	D	E
1	Truck Rental Markov Model				
2					
3	From Birmingham to Columbia =				25.0%
4	From Birmingham to Dallas =				25.0%
5	From Columbia to Birmingham =				20.0%
6	From Columbia to Dallas =				40.0%
7	From Dallas to Birmingham =				15.0%
8	From Dallas to Columbia =				5.0%
9					
10	t	Birmingham	Columbia	Dallas	
11	0	200	200	200	
12	1	170.00	140.00	290.00	
25	14	145.46	90.91	363.63	
26	15	145.45	90.91	363.64	
27	16	145.45	90.91	363.64	

FIGURE 7.3 Excel output for Exercise 7.1.1c.

d. Use Excel to determine the long-term distribution of trucks in each city.

Finding the long-term distribution of trucks in each city means we have to find the long-term percentages of trucks in each city. Following Example 7.2, we add columns to our spreadsheet where we keep track of these percentages. Then we copy the formulas down until the percentages no longer change. The result is shown in Figure 7.4. We end up with 24.24% of the fleet in Birmingham, 15.15% in Columbia, and 60.61% in Dallas.

	A	B	C	D	E	F	G
1	Truck Rental Markov Model						
2							
3	From Birmingham to Columbia =				25.0%		
4	From Birmingham to Dallas =				25.0%		
5	From Columbia to Birmingham =				20.0%		
6	From Columbia to Dallas =				40.0%		
7	From Dallas to Birmingham =				15.0%		
8	From Dallas to Columbia =				5.0%		
9							
10	t	Birmingham	Columbia	Dallas	Birm. %	Col. %	Dal. %
11	0	200	200	200	33.33%	33.33%	33.33%
12	1	170.00	140.00	290.00	28.33%	23.33%	48.33%
25	14	145.46	90.91	363.63	24.24%	15.15%	60.61%
26	15	145.45	90.91	363.64	24.24%	15.15%	60.61%
27	16	145.45	90.91	363.64	24.24%	15.15%	60.61%

FIGURE 7.4 Excel output for Exercise 7.1.1d.

3 Suppose a truck rental company has locations in Birmingham, Alabama; Columbia, South Carolina; Dallas, Texas; and Evansville, Indiana. The company permits one-way rentals and knows based on past experience that during a typical week the following truck movements occur.

Of the trucks that start in Birmingham:
- 50% stay in Birmingham;
- 25% travel to Columbia;
- 15% travel to Dallas; and
- 10% travel to Evansville.

Of the trucks that start in Columbia:
- 40% stay in Columbia;
- 14% travel to Birmingham;
- 34% travel to Dallas; and
- 12% travel to Evansville.

Of the trucks that start in Dallas:
- 75% stay in Dallas;
- 15% travel to Birmingham;
- 5% travel to Columbia; and
- 5% travel to Evansville.

Of the trucks that start in Evansville:
- 65% stay in Evansville;
- 10% travel to Birmingham;
- 10% travel to Columbia; and
- 15% travel to Dallas.

a. Give a flow diagram for the model.

The flow diagram is given in Figure 7.5.

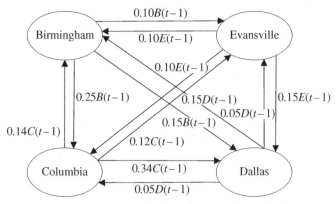

FIGURE 7.5   Flow diagram for Exercise 7.1.3.

b. Implement the model in Excel.

First we give the DDS:

$$B(t) = B(t-1) - 0.25B(t-1) - 0.15B(t-1) - 0.10B(t-1)$$
$$+ 0.14C(t-1) + 0.15D(t-1) + 0.10E(t-1)$$

$$C(t) = C(t-1) - 0.14C(t-1) - 0.34C(t-1) - 0.12C(t-1)$$
$$+ 0.25B(t-1) + 0.05D(t-1) + 0.10E(t-1)$$

$$D(t) = D(t-1) - 0.15D(t-1) - 0.05D(t-1) - 0.05D(t-1)$$
$$+ 0.15B(t-1) + 0.34C(t-1) + 0.15E(t-1)$$

$$E(t) = E(t-1) - 0.10E(t-1) - 0.10E(t-1) - 0.15E(t-1)$$
$$+ 0.10B(t-1) + 0.12C(t-1) + 0.05D(t-1).$$

Next we add to the 3-city Excel spreadsheet. We need to insert new rows at the top to allow space for the new parameters, and we need to add a column for Evansville. The set-up with the formula for Birmingham displayed is given in Figure 7.6.

	A	B	C	D	E	F	G	H
1	Truck Rental Markov Model							
2								
3	From Birmingham to Columbia =				25.0%			
4	From Birmingham to Dallas =				15.0%			
5	From Birmingham to Evansville =				10.0%			
6	From Columbia to Birmingham =				14.0%			
7	From Columbia to Dallas =				34.0%			
8	From Columbia to Evansville =				12.0%			
9	From Dallas to Birmingham =				15.0%			
10	From Dallas to Columbia =				5.0%			
11	From Dallas to Evansville =				5.0%			
12	From Evansville to Birmingham =				10.0%			
13	From Evansville to Columbia =				10.0%			
14	From Evansville to Dallas =				15.0%			
15								
16	t	Birmingham	Columbia	Dallas	Evansville			
17	0	200	200	200	200			
18	1	=B17-$E$3*B17-$E$4*B17-$E$5*B17+$E$6*C17+$E$9*D17+$E$12*E17						

FIGURE 7.6 Excel set-up for Exercise 7.1.3.

c. If 200 trucks begin in each city, use Excel to determine the long-term numbers of trucks in each city.

We copy the formulas down until the number of trucks in each city stops changing. Figure 7.7 shows the result with most rows hidden. We end up with 172 trucks in Birmingham, 125 in Columbia, 360 in Dallas, and 143 in Evansville.

d. Use Excel to determine the long-term distribution of trucks in each city.

We need to add columns where we have Excel calculate the percentage of trucks in each city. For example, the percentage of trucks in Evansville is found by calculating

	A	B	C	D	E
1	Truck Rental Markov Model				
2					
3	From Birmingham to Columbia =				25.0%
4	From Birmingham to Dallas =				15.0%
5	From Birmingham to Evansville =				10.0%
6	From Columbia to Birmingham =				14.0%
7	From Columbia to Dallas =				34.0%
8	From Columbia to Evansville =				12.0%
9	From Dallas to Birmingham =				15.0%
10	From Dallas to Columbia =				5.0%
11	From Dallas to Evansville =				5.0%
12	From Evansville to Birmingham =				10.0%
13	From Evansville to Columbia =				10.0%
14	From Evansville to Dallas =				15.0%
15					
16	t	Birmingham	Columbia	Dallas	Evansville
17	0	200	200	200	200
18	1	178.00	160.00	278.00	184.00
37	20	171.66	125.39	359.56	143.40
38	21	171.66	125.39	359.56	143.40

FIGURE 7.7   Excel output for Exercise 7.1.3c.

$$\frac{E(t)}{B(t) + C(t) + D(t) + E(t)}.$$

Figure 7.8 shows the Excel set-up, the results, and the formula for the percentage of trucks in Birmingham displayed. We end up with 21.5% of the fleet in Birmingham, 15.7% in Columbia, 44.9% in Dallas, and 17.9% in Evansville.

	A	B	C	D	E	F	G	H	I
1	Truck Rental Markov Model								
2									
3	From Birmingham to Columbia =				25.0%				
4	From Birmingham to Dallas =				15.0%				
5	From Birmingham to Evansville =				10.0%				
6	From Columbia to Birmingham =				14.0%				
7	From Columbia to Dallas =				34.0%				
8	From Columbia to Evansville =				12.0%				
9	From Dallas to Birmingham =				15.0%				
10	From Dallas to Columbia =				5.0%				
11	From Dallas to Evansville =				5.0%				
12	From Evansville to Birmingham =				10.0%				
13	From Evansville to Columbia =				10.0%				
14	From Evansville to Dallas =				15.0%				
15									
16	t	Birmingham	Columbia	Dallas	Evansville	Birm. %	Col. %	Dal. %	Evan. %
17	0	200	200	200	200	25.0%	25.0%	25.0%	25.0%
18	1	178.00	160.00	278.00	184.00	=B18/($B$17+$C$17+$D$17+$E$17)			
37	20	171.66	125.39	359.56	143.40	21.5%	15.7%	44.9%	17.9%
38	21	171.66	125.39	359.56	143.40	21.5%	15.7%	44.9%	17.9%

FIGURE 7.8   Excel output for Exercise 7.1.3d.

5 For the situation in Exercise 7.1.1, find the largest fleet the company can accommodate if it can house 100 trucks in Birmingham, 50 trucks in Columbia, and 150 trucks in Dallas.

Recall that the long-term distribution of trucks for Exercise 7.1.1 was 24.24% of the fleet in Birmingham, 15.15% in Columbia, and 60.61% in Dallas.
   If Birmingham can only house 100 trucks, then the largest fleet it can serve is

$$.2424 \times N = 100$$
$$N \approx 412.5.$$

If Columbia can only house 50 trucks, then the largest fleet it can serve is

$$.1515 \times N = 50$$
$$N \approx 330.03.$$

If Dallas can only house 150 trucks, then the largest fleet it can serve is

$$.6061 \times N = 150$$
$$N \approx 247.48.$$

The largest possible fleet will be the smallest fleet any one city can serve. In this case the largest fleet will be 247 trucks. Dallas is the limiting city.

7 Considering the flow diagram in Exercise 7.1.1, how do we know that we will end up with a stable distribution?

The flow diagram is irreducible since trucks may travel from any city to any other city. The flow diagram is also aperiodic because it is possible for a truck to remain in any given city. By Theorem 7.1 we know we will get a positive, stable distribution.

9 Give an example of a 4-city flow diagram for which there will be no positive stable distribution. What condition(s) of Theorem 7.1 does the diagram violate?

Such a diagram is given in Figure 7.9. The flow diagram is irreducible because it is possible for a truck to travel between any two cities (though not necessarily in one step). The flow diagram is not aperiodic: in order to return to any given city, a truck must move a number of steps that is a multiple of 4. Thus each city has period 4.

11 Interpret the flow diagram for Exercise 7.1.1 from a random walk point of view.

If an individual truck is in Birmingham, then it has a 50% chance of staying in Birmingham, a 25% chance of moving to Columbia, and a 25% chance of moving to Dallas the following week. We view the other cities the same way.
   We view the stable distribution as the percentage of time an individual truck will spend in each city. Thus any particular truck will spend 24.24% of

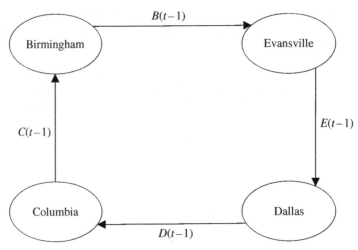

FIGURE 7.9   Flow diagram for Exercise 7.1.9.

its weeks in Birmingham, 15.15% of its weeks in Columbia, and 60.61% of its weeks in Dallas.

13   Give an example of a flow diagram with 6 states that is neither irreducible nor aperiodic.

Figure 7.10 shows such a diagram. It is not possible to travel to State 1 so the diagram is not irreducible. State 5 has period 3, so the diagram is not aperiodic.

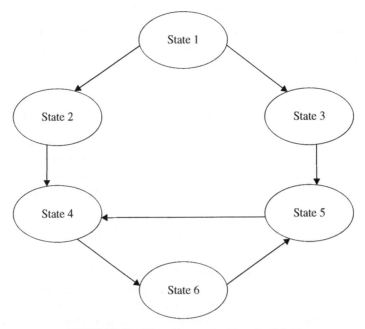

FIGURE 7.10   Flow diagram for Exercise 7.1.13.

7.2   RANKING SPORTS TEAMS

1   Text Table 7.17 gives the results of a season of games with no undefeated teams. Rank the teams using the Markov method from Example 7.8. Discuss any rankings that differ from your expectations.

TEXT TABLE 7.17   Complete Wins and Losses for 6-Team League

Teams	Bears	Cardinals	Dolphins	Eagles	Falcons	Giants
Bears	0	1	1	0	1	1
Cardinals	0	0	1	1	0	1
Dolphins	0	0	0	1	1	1
Eagles	1	0	0	0	1	0
Falcons	0	1	0	0	0	0
Giants	0	0	0	1	1	0

We begin by using Text Table 7.17 to compile the records of each team, which we present in Table 7.1.

TABLE 7.1   Team Records for 6-Team League

Team	Wins	Losses
Bears	4	1
Cardinals	3	2
Dolphins	3	2
Eagles	2	3
Falcons	1	4
Giants	2	3

Next we create a table that shows the proportions of losing teams' credit that is distributed to the winning team. Our convention is that a team with L losses will distribute $\frac{1}{L+2}$ of its credit to each team that defeated it. For example, here the Eagles, who lost 3 games, will distribute $\frac{1}{L+2} = \frac{1}{3+2} = \frac{1}{5}$ of their credit to each of the teams that beat them. We get Table 7.2.

TABLE 7.2   Proportion of Losing Team's Credit Given to Winning Team

Teams	Bears	Cardinals	Dolphins	Eagles	Falcons	Giants
Bears	0	1/4	1/4	0	1/6	1/5
Cardinals	0	0	1/4	1/5	0	1/5
Dolphins	0	0	0	1/5	1/6	1/5
Eagles	1/3	0	0	0	1/6	0
Falcons	0	1/4	0	0	0	0
Giants	0	0	0	1/5	1/6	0

Based on Table 7.2 we form the DDS for the model:

$$B(t) = B(t-1) - \frac{1}{3}B(t-1) + \frac{1}{4}C(t-1) + \frac{1}{4}D(t-1) + \frac{1}{6}F(t-1) + \frac{1}{5}G(t-1)$$

$$C(t) = C(t-1) - \frac{1}{2}C(t-1) + \frac{1}{4}D(t-1) + \frac{1}{5}E(t-1) + \frac{1}{5}G(t-1)$$

$$D(t) = D(t-1) - \frac{1}{2}D(t-1) + \frac{1}{5}E(t-1) + \frac{1}{6}F(t-1) + \frac{1}{5}G(t-1)$$

$$E(t) = E(t-1) - \frac{3}{5}E(t-1) + \frac{1}{3}B(t-1) + \frac{1}{6}F(t-1)$$

$$F(t) = F(t-1) - \frac{2}{3}F(t-1) + \frac{1}{4}C(t-1)$$

$$G(t) = G(t-1) - \frac{3}{5}G(t-1) + \frac{1}{5}E(t-1) + \frac{1}{6}F(t-1).$$

Entering the DDS correctly in Excel takes some time. We show the set-up with the formula for the Bears displayed in Figure 7.11. Since we know we will get a positive stable distribution, the initial amount of credit is unimportant. We assign each team 1 unit of credit initially.

	A	B	C	D	E	F	G
1	Football Team Rankings						
2							
3	t	Bears	Cardinals	Dolphins	Eagles	Falcons	Giants
4	0	1	1	1	1	1	1
5	1	=B4-(1/3)*B4+(1/4)*C4+(1/4)*D4+(1/6)*F4+(1/5)*G4					0.77
6	2	1.83	1.18	0.96	0.97	0.48	0.58
7	3	1.95	1.14	0.87	1.08	0.45	0.51

FIGURE 7.11   Excel set-up for Exercise 7.2.1.

To finish the ranking we add columns where we get Excel to calculate the percentage of credit for each team at each time step. Figure 7.12 shows the formula for the Bears' percentage.

	H	I	J	K	L	M
1						
2						
3	B%	C%	D%	E%	F%	G%
4	=B4/($B$4+$C$4+$D$4+$E$4+$F$4+$G$4)					16.67%
5	25.56%	19.17%	17.78%	15.00%	9.72%	12.78%
6	30.45%	19.58%	16.06%	16.14%	8.03%	9.73%

FIGURE 7.12   Excel set-up with percentages for Exercise 7.2.1.

Finally, we copy the formulas down until the percentages of credit are no longer changing, and we rank teams from most to least credit. In this example it takes about 20 iterations, or steps, in the model to arrive at the stable distribution. The ranked results are presented in Table 7.3.

TABLE 7.3   Markov Method Rankings for 6-Team League

Team	Credit (%)
Bears	32.57
Eagles	20.00
Cardinals	18.29
Dolphins	13.71
Giants	8.57
Falcons	6.86

Note that the Bears, who had the best record, are ranked first, and the Falcons, who had the worst record are ranked last. The Eagles, however, who had a 2–3 record are ranked above the two teams that were 3–2. This is because the Eagles received a lot of credit for being the only team to defeat the Bears.

3   If teams do not all play each other it is possible to have more than one undefeated team, and when we have more than one undefeated team there is more than one way we can proceed with a ranking. One way is to use the just-for-playing credit, but here we look at another option similar to Example 7.9: instead of just-for-playing credit we can first use our original ranking method on all teams, then we remove the undefeated teams and re-run the method. Text Table 7.19 gives the results of a season of games with two undefeated teams.

TEXT TABLE 7.19   Complete Wins and Losses for 6-Team League

Teams	Bears	Cardinals	Dolphins	Eagles	Falcons	Giants
Bears	0	1	1	1	0	0
Cardinals	0	0	0	1	1	0
Dolphins	0	0	0	0	1	0
Eagles	0	0	0	0	0	0
Falcons	0	0	0	0	0	0
Giants	0	0	1	1	1	0

a. Rank the teams without the just-for-playing credit.

We begin by using Text Table 7.19 to compile the records of each team, which we present in Table 7.4.

Next we create a table that shows the proportions of losing teams' credit that is distributed to the winning team. Our convention is that a team with L losses will distribute $\frac{1}{L+2}$ of its credit to each team that defeated it. For example, here the Eagles, who lost 3 games, will distribute $\frac{1}{L+2} = \frac{1}{3+2} = \frac{1}{5}$ of their credit to each of the teams that beat them. We get Table 7.5.

TABLE 7.4   Team Records for 6-Team League

Team	Wins	Losses
Bears	3	0
Cardinals	2	1
Dolphins	1	2
Eagles	0	3
Falcons	0	3
Giants	3	0

TABLE 7.5   Proportion of Losing Team's Credit Given to Winning Team

Teams	Bears	Cardinals	Dolphins	Eagles	Falcons	Giants
Bears	0	1/3	1/4	1/5	0	0
Cardinals	0	0	0	1/5	1/5	0
Dolphins	0	0	0	0	1/5	0
Eagles	0	0	0	0	0	0
Falcons	0	0	0	0	0	0
Giants	0	0	1/4	1/5	1/5	0

Based on Table 7.5 we form the DDS for the model:

$$B(t) = B(t-1) + \frac{1}{3}C(t-1) + \frac{1}{4}D(t-1) + \frac{1}{5}E(t-1)$$

$$C(t) = C(t-1) - \frac{1}{3}C(t-1) + \frac{1}{5}E(t-1) + \frac{1}{5}F(t-1)$$

$$D(t) = D(t-1) - \frac{1}{2}D(t-1) + \frac{1}{5}F(t-1)$$

$$E(t) = E(t-1) - \frac{3}{5}E(t-1)$$

$$F(t) = F(t-1) - \frac{3}{5}F(t-1)$$

$$G(t) = G(t-1) + \frac{1}{4}D(t-1) + \frac{1}{5}E(t-1) + \frac{1}{5}F(t-1).$$

We show the Excel set-up with the formula for the Bears displayed in Figure 7.13. Since all teams should be considered to be on equal footing pre-season, we assign each team 1 unit of credit initially.

	A	B	C	D	E	F	G
1	Football Team Rankings						
2							
3	t	Bears	Cardinals	Dolphins	Eagles	Falcons	Giants
4	0	1	1	1	1	1	1
5	1	=B4+(1/3)*C4+(1/4)*D4+(1/5)*E4			0.40	0.40	1.65
6	2	2.39	0.87	0.43	0.16	0.16	1.99
7	3	2.82	0.64	0.25	0.06	0.06	2.16

FIGURE 7.13   Excel set-up for Exercise 7.2.3.

To finish the ranking we add columns where we get Excel to calculate the percentage of credit for each team at each time step. Figure 7.14 shows the formula for the Bears' percentage.

	H	I	J	K	L	M
1						
2						
3	B%	C%	D%	E%	F%	G%
4	=B4/($B$4+$C$4+$D$4+$E$4+$F$4+$G$4)					16.67%
5	29.72%	17.78%	11.67%	6.67%	6.67%	27.50%
6	39.90%	14.52%	7.17%	2.67%	2.67%	33.08%

FIGURE 7.14 Excel set-up with percentages for Exercise 7.2.3.

Finally, we copy the formulas down until the percentages of credit are no longer changing, and we rank teams from most to least credit. The ranked results are presented in Table 7.6.

TABLE 7.6 Markov Method Rankings for 6-Team League

Team	Credit (%)
Bears	61.11
Giants	38.89
Cardinals	0
Dolphins	0
Eagles	0
Falcons	0

b. Interpret the results of the ranking.

All of the credit has ended up with the two undefeated teams, the Bears and the Giants, so the ranking is not useful for the remaining four teams. However, the method does differentiate between the two undefeated teams: the Bears are ranked higher. The reason is that the Bears had the better wins.

c. After removing the undefeated teams, re-run the method.

If we remove the undefeated Bears and Giants, then the table of wins and losses is given in Table 7.7.

TABLE 7.7 Complete Wins and Losses for 4-Team League

Teams	Cardinals	Dolphins	Eagles	Falcons
Cardinals	0	0	1	1
Dolphins	0	0	0	1
Eagles	0	0	0	0
Falcons	0	0	0	0

Then Table 7.8 shows how credit will be distributed. The ranking carried out on the four remaining teams results in Table 7.9. We have now differentiated

TABLE 7.8 **The Distribution of Credit for Remaining 4 Teams**

Teams	Cardinals	Dolphins	Eagles	Falcons
Cardinals	0	0	1/3	1/4
Dolphins	0	0	0	1/4
Eagles	0	0	0	0
Falcons	0	0	0	0

TABLE 7.9 **Credit Distribution Among Remaining 4 Teams**

Team	Credit (%)
Cardinals	62.5
Dolphins	37.5
Eagles	0
Falcons	0

between the Cardinals and the Dolphins, but we still have not differentiated between the winless Eagles and the Falcons. At this point we can go no further with the original method. We can, however, note that the Falcons had the worst loss in losing to the 1-2 Dolphins, so we can reasonably rank the Eagles ahead of them.

d. Compare the final ranking of this method with the ranking produced by the just-for-playing credit.

The final ranking for the original method is:

1. Bears
2. Giants
3. Cardinals
4. Dolphins
5. Eagles
6. Falcons.

If we use the just-for-playing credit method for ranking the original 6 teams the table of credit is given in Table 7.10.

TABLE 7.10 **How Credit is Distributed with Just-for-Playing Credit**

Teams	Bears	Cardinals	Dolphins	Eagles	Falcons	Giants
Bears	0	1/3	1/4	1/5	0	0
Cardinals	1/36	0	0	1/5	1/5	0
Dolphins	1/36	0	0	0	1/5	1/36
Eagles	1/36	1/36	0	0	0	1/36
Falcons	0	1/36	1/36	0	0	1/36
Giants	0	0	1/4	1/5	1/5	0

The DDS for the system is then

$$B(t) = B(t-1) - \frac{1}{12}B(t-1) + \frac{1}{3}C(t-1) + \frac{1}{4}D(t-1) + \frac{1}{5}E(t-1)$$

$$C(t) = C(t-1) - \frac{14}{36}C(t-1) + \frac{1}{36}B(t-1) + \frac{1}{5}E(t-1) + \frac{1}{5}F(t-1)$$

$$D(t) = D(t-1) - \frac{19}{36}D(t-1) + \frac{1}{36}B(t-1) + \frac{1}{5}F(t-1) + \frac{1}{36}G(t-1)$$

$$E(t) = E(t-1) - \frac{3}{5}E(t-1) + \frac{1}{36}B(t-1) + \frac{1}{36}C(t-1) + \frac{1}{36}G(t-1)$$

$$F(t) = F(t-1) - \frac{3}{5}F(t-1) + \frac{1}{36}C(t-1) + \frac{1}{36}D(t-1) + \frac{1}{36}G(t-1)$$

$$G(t) = G(t-1) - \frac{1}{12}G(t-1) + \frac{1}{4}D(t-1) + \frac{1}{5}E(t-1) + \frac{1}{5}F(t-1).$$

After implementing the system in Excel, the resulting ranking is given in Table 7.11.

TABLE 7.11   Just-for-Playing Markov Ranking for 6-Team League

Team	Credit (%)
Bears	52.30
Giants	29.76
Cardinals	6.84
Dolphins	5.05
Eagles	4.12
Falcons	1.93

The ranking using the just-for-playing credit is the same as the one generated in parts a—c.

5   Text Table 7.21 gives the 2014 NCAA football season results for the SEC East.

TEXT TABLE 7.21   Complete Wins and Losses for 2014 SEC East

Teams	Florida	Georgia	Kentucky	Missouri	South Carolina	Tennessee	Vanderbilt
Florida	0	1	1	0	0	1	1
Georgia	0	0	1	1	0	1	1
Kentucky	0	0	0	0	1	0	1
Missouri	1	0	1	0	1	1	1
South Carolina	1	1	0	0	0	0	1
Tennessee	0	0	1	0	1	0	1
Vanderbilt	0	0	0	0	0	0	0

Only games between Eastern division teams are included. Rank the teams. Discuss any ranking that differs from your expectations.

The team records are given in Table 7.12.

Because there is a winless team we use the just-for-playing ranking method. Table 7.13 shows how credit will be distributed.

TABLE 7.12 Team Records for 2014 SEC East

Team	Wins	Losses
Florida	4	2
Georgia	4	2
Kentucky	2	4
Missouri	5	1
South Carolina	3	3
Tennessee	3	3
Vanderbilt	0	6

TABLE 7.13 Just-for-Playing Credit Distribution for 2014 SEC East

Teams	Florida	Georgia	Kentucky	Missouri	South Carolina	Tennessee	Vanderbilt
Florida	0	1/4	1/6	1/49	1/49	1/5	1/8
Georgia	1/49	0	1/6	1/3	1/49	1/5	1/8
Kentucky	1/49	1/49	0	1/49	1/5	1/49	1/8
Missouri	1/4	1/49	1/6	0	1/5	1/5	1/8
South Carolina	1/4	1/4	1/49	1/49	0	1/49	1/8
Tennessee	1/49	1/49	1/6	1/49	1/5	0	1/8
Vanderbilt	1/49	1/49	1/49	1/49	1/49	1/49	0

The corresponding DDS with the dependence on time suppressed is given by

$$F = F - \frac{57}{98}F + \frac{1}{4}G + \frac{1}{6}K + \frac{1}{49}M + \frac{1}{49}SC + \frac{1}{5}T + \frac{1}{8}V$$

$$G = G - \frac{57}{98} + \frac{1}{49}F + \frac{1}{6}K + \frac{1}{3}M + \frac{1}{49}SC + \frac{1}{5}T + \frac{1}{8}V$$

$$K = K - \frac{104}{147}K + \frac{1}{49}F + \frac{1}{49}G + \frac{1}{49}M + \frac{1}{5}SC + \frac{1}{49}T + \frac{1}{8}V$$

$$M = M - \frac{64}{147}M + \frac{1}{4}F + \frac{1}{49}G + \frac{1}{6}K + \frac{1}{5}SC + \frac{1}{5}T + \frac{1}{8}V$$

$$SC = SC - \frac{162}{245}SC + \frac{1}{4}F + \frac{1}{4}G + \frac{1}{49}K + \frac{1}{49}M + \frac{1}{49}T + \frac{1}{8}V$$

$$T = T - \frac{162}{245}T + \frac{1}{49}F + \frac{1}{49}G + \frac{1}{6}K + \frac{1}{49}M + \frac{1}{5}SC + \frac{1}{8}V$$

$$V = V - \frac{3}{4}V + \frac{1}{49}F + \frac{1}{49}G + \frac{1}{49}K + \frac{1}{49}M + \frac{1}{49}SC + \frac{1}{49}T.$$

After implementing the DDS in Excel, calculating percentages of credit for each team, and dragging the formulas down until the percentages no longer change, we get the ranking shown in Table 7.14.

TABLE 7.14 Markov Ranking for 2014 SEC East

Team	Credit (%)
Missouri	25.93
Georgia	21.85
Florida	16.70
South Carolina	16.38
Tennessee	9.27
Kentucky	7.23
Vanderbilt	2.65

There does not seem to be anything controversial in the ranking. It agrees with the ranking one would produce using won-lost records, and it has broken the ties that would exist between Florida and Georgia and between South Carolina and Tennessee.

7 *Extension*: For a sport and league or conference of your choosing, rank the teams with the Markov just-for-playing method.

Outside of finding the necessary data to generate the table of wins and losses, there is no additional difficulty in this one. If the number of teams is large, the problem can become labor intensive.

9 *Extension*: Discuss how to use a Markov model to rank the importance of words in the English language.

The idea here is to create a flow diagram for the entire English language. Each word gets its own oval, and arrows extend from one word to another if the destination word is used in the definition of originating word. In this way words distribute importance to other words used in their definition similar to the way a sports team distributes credit to those that beat it. Words that are used in the definitions of many other words receive a lot of importance. Words that require a lot of words in their definition distribute a lot their importance to others. An "undefeated" word would be one that distributed no importance to others, but this would mean a word with no definition at all so we can rule this situation out. A "winless" word would be a word that is not used in any other word's definition. We would not want to assign 0 importance to such a word, and we could avoid doing so with an adjustment akin to the "just-for-playing" credit.

7.3 GOOGLE PAGERANK

1 Determine whether the web represented in Text Figure 7.23 contains any dangling nodes or sub-webs.

The web in Text Figure 7.23 contains neither. All pages have links to others so there are no dangling nodes, and it is possible to get from any page to any other so there are no sub-webs.

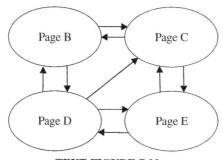

TEXT FIGURE 7.23

3 Determine whether the web represented in Text Figure 7.25 contains any dangling nodes or sub-webs.

Page E is a dangling node because it has no outgoing links. Pages B, D, and E form a sub-web because once a web-surfer is at one of those three pages, the surfer cannot leave them to get back to Page C.

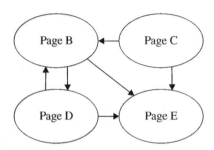

TEXT FIGURE 7.25

5 Consider the web represented in Text Figure 7.23.

    a. Label the arrows with the proportions of importance that each origination page distributes to the destination page.

The labeled web diagram is shown in Figure 7.15.

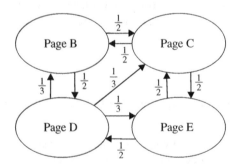

FIGURE 7.15   Web diagram for Exercise 7.3.5.

    b. Create a hyperlink table from the flow diagram.

Table 7.15 shows the hyperlink table.

TABLE 7.15   **Hyperlink Table for Exercise 7.3.5**

HT	B	C	D	E
B	0	1/2	1/3	0
C	1/2	0	1/3	1/2
D	1/2	0	0	1/2
E	0	1/2	1/3	0

c. Fix any dangling nodes by creating the modified hyperlink table.

Since there are no dangling nodes it is not necessary to create the modified hyperlink table.

d. Form the teleportation table.

Table 7.16 shows the teleportation table.

TABLE 7.16   Teleportation Table for Exercise 7.3.5

T	B	C	D	E
B	1/4	1/4	1/4	1/4
C	1/4	1/4	1/4	1/4
D	1/4	1/4	1/4	1/4
E	1/4	1/4	1/4	1/4

e. Fix any sub-webs by forming the Google table.

The Google table is given in Table 7.17.

TABLE 7.17   Google Table for Exercise 7.3.5

$GT = \frac{17}{20} \times HT + \frac{3}{20} \times T$	B	C	D	E
B	$\frac{3}{80}$	$\frac{37}{80}$	$\frac{77}{240}$	$\frac{3}{80}$
C	$\frac{37}{80}$	$\frac{3}{80}$	$\frac{77}{240}$	$\frac{37}{80}$
D	$\frac{37}{80}$	$\frac{3}{80}$	$\frac{3}{80}$	$\frac{37}{80}$
E	$\frac{3}{80}$	$\frac{37}{80}$	$\frac{77}{240}$	$\frac{3}{80}$

f. Rank the webpages in order of importance.

The DDS for the Google table with the dependence on time suppressed is given by

$$B(t) = \frac{3}{80}B + \frac{37}{80}C + \frac{77}{240}D + \frac{3}{80}E$$

$$C(t) = \frac{37}{80}B + \frac{3}{80}C + \frac{77}{240}D + \frac{37}{80}E$$

$$D(t) = \frac{37}{80}B + \frac{3}{80}C + \frac{3}{80}D + \frac{37}{80}E$$

$$E(t) = \frac{3}{80}B + \frac{37}{80}C + \frac{77}{240}D + \frac{3}{80}E.$$

Implementing the DDS in Excel is a matter of being careful when typing in the formulas. Knowing that the Google ranking method always produces a positive stable distribution allows us to start with any amount of initial importance, so for convenience all pages start with importance equal to one. We show the Excel set-up in Figure 7.16. The resulting ranking is given in Table 7.18.

	A	B	C	D	E	F	G	H	I
1	Web Page Rankings								
2									
3	*t*	*B*	*C*	*D*	*E*	*B%*	*C%*	*D%*	*E%*
4	0	1	1	1	1	25.00%	25.00%	25.00%	25.00%
5	1	=(3/80)*B4+(37/80)*C4+(77/240)*D4+(3/80)*E4					32.08%	25.00%	21.46%
6	2	0.98	1.16	0.88	0.98	24.47%	29.07%	21.99%	24.47%

FIGURE 7.16   Excel set-up for Exercise 7.3.5.

TABLE 7.18   Ranking of Importance by Google Table Exercise 7.3.5

Webpage	Importance (%)
C	30.12
D	23.47
B	23.20
E	23.20

7   Consider the web represented in Text Figure 7.25.

   a. Label the arrows with the proportions of importance that each origination page distributes to the destination page.

   The labeled web diagram is shown in Figure 7.17.

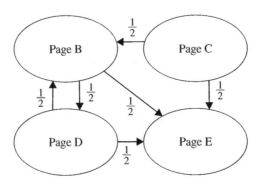

FIGURE 7.17   Web diagram for Exercise 7.3.7.

   b. Create a hyperlink table from the flow diagram.

   Table 7.19 shows the hyperlink table.

TABLE 7.19   Hyperlink Table for Exercise 7.3.7

HT	B	C	D	E
B	0	1/2	1/2	0
C	0	0	0	0
D	1/2	0	0	0
E	1/2	1/2	1/2	0

c. Fix any dangling nodes by creating the modified hyperlink table.

Page E is a dangling node so we need to form the modified hyperlink table. It is given in Table 7.20.

TABLE 7.20   Modified Hyperlink Table for Exercise 7.3.7

HT'	B	C	D	E
B	0	1/2	1/2	1/4
C	0	0	0	1/4
D	1/2	0	0	1/4
E	1/2	1/2	1/2	1/4

d. Form the teleportation table.

Table 7.21 shows the teleportation table.

TABLE 7.21   Teleportation Table for Exercise 7.3.7

T	B	C	D	E
B	1/4	1/4	1/4	1/4
C	1/4	1/4	1/4	1/4
D	1/4	1/4	1/4	1/4
E	1/4	1/4	1/4	1/4

e. Fix any sub-webs by forming the Google table.

The Google table is given in Table 7.22.

TABLE 7.22   Google Table for Exercise 7.3.7

$GT = \frac{17}{20} \times HT + \frac{3}{20} \times T$	B	C	D	E
B	$\frac{3}{80}$	$\frac{37}{80}$	$\frac{37}{80}$	$\frac{1}{4}$
C	$\frac{3}{80}$	$\frac{3}{80}$	$\frac{3}{80}$	$\frac{1}{4}$
D	$\frac{37}{80}$	$\frac{3}{80}$	$\frac{3}{80}$	$\frac{1}{4}$
E	$\frac{37}{80}$	$\frac{37}{80}$	$\frac{37}{80}$	$\frac{1}{4}$

f. Rank the webpages in order of importance.

The DDS for the Google table with the dependence on time suppressed is given by

$$B(t) = \frac{3}{80}B + \frac{37}{80}C + \frac{37}{80}D + \frac{1}{4}E$$

$$C(t) = \frac{3}{80}B + \frac{3}{80}C + \frac{3}{80}D + \frac{1}{4}E$$

$$D(t) = \frac{37}{80}B + \frac{3}{80}C + \frac{3}{80}D + \frac{1}{4}E$$

$$E(t) = \frac{37}{80}B + \frac{37}{80}C + \frac{37}{80}D + \frac{1}{4}E.$$

TABLE 7.23   Ranking of Importance by Google Table Exercise 7.3.7

Webpage	Importance (%)
E	38.14
B	26.77
D	23.23
C	11.86

Implementing the DDS in Excel is a matter of being careful when typing in the formulas. Knowing that the Google ranking method always produces a positive stable distribution allows us to start with any amount of initial importance, so for convenience all pages start with importance equal to one. The resulting ranking is given in Table 7.23.

8

BODY WEIGHT AND BODY COMPOSITION

8.1 CONSTANT CALORIE EXPENDITURE

1 Implement the constant calorie expenditure model with Excel using your own body weight and calorie parameters. Note what the model predicts for your body weight over 1 week, 1 month, 6 months, 1 year, and 5 years. At what point does the model start to become unreasonable?

For example purposes we assume a body weight of 170 pounds, a daily calorie intake of $I_0 = 2,000$, and a daily calorie expenditure of $E_0 = 2,200$. We input the parameters into the constant calorie expenditure Excel spreadsheet, and copy the formulas down to 5 years, or $t = 1,825$ days. By hiding most of the rows we can record all required body weights in Figure 8.1.

There is not one specific day on which the projected body weight becomes unreasonable, but it certainly looks like the projection for year 5 (65.7 pounds) is too low. Sometime between year 1 and year 5 the projection starts to become unreasonable.

3 Suppose someone consumes 100 calories per day more than they burn. How much weight will the person gain over a 2 week period?

Regardless of the person's initial weight, a 100 calorie per day calorie surplus over a 2-week period results in $14 \times 100 = 1,400$ total excess calories. Using the estimate that 1 pound of body weight is equivalent to 3,500 calories allows us to determine that the total weight gain will be $\frac{1,400}{3,500} = 0.40$ pounds.

Solutions Manual to Accompany Models for Life: An Introduction to Discrete Mathematical Modeling with Microsoft® Office Excel®, First Edition. Jeffrey T. Barton.
© 2016 John Wiley & Sons, Inc. Published 2016 by John Wiley & Sons, Inc.
Companion website: www.wiley.com/go/barton/solutionsmanual_modelsforlife

	A	B	C	D
1	Constant Calorie Expenditure Model			
2				
3	Current Body Weight =		170	(pounds)
4	Daily Intake, I_0 =		2000	(calories)
5	Daily Expenditure, E_0 =		2200	(calories)
6				
7	t	$W(t)$		
8	0	170		
9	1	169.9		
15	7	169.6	1 week	
38	30	168.3	1 month	
191	183	159.5	6 months	
373	365	149.1	1 year	
1833	1825	65.7	5 years	

FIGURE 8.1   Excel output for Exercise 8.1.1.

5   How long will it take to lose 10 pounds with a calorie deficit of 1 calorie per day?

To lose 10 pounds requires a total calorie deficit of $10 \times 3,500 = 35,000$ calories. With a deficit of 1 calorie per day, this will take 35,000 days, or about $\frac{35,000}{365} \approx 96$ years.

8.2   VARIABLE CALORIE EXPENDITURE

1   Implement the variable calorie expenditure model with Excel using your own parameters for initial weight, height, age, activity level, and sex, and select a daily calorie intake that would result in weight gain.

a. Note what the model predicts for your body weight over 1 week, 1 month, 6 months, 1 year, and 5 years.

For example purposes we assume a 28-year-old male who is 68 inches tall and weighs 170 pounds. We assume an activity level of moderate and a daily calorie intake of $I_0 = 3,000$. We input the parameters into the variable calorie expenditure Excel spreadsheet, and copy the formulas down to 5 years, or $t = 1,825$ days. By hiding most of the rows we can record all required body weights in Figure 8.2.

b. Use Excel to determine your long-term stable weight.

We copy the formulas down until the body weight stops changing. In this example the weight levels off at about 218.4 pounds.

3   For the parameters you used in Exercise 8.2.1, determine the long-term weight by finding the equilibrium value algebraically.

	A	B	C	D
1	Variable Calorie Expenditure Model			
2				
3	Sex, S =		M	
4	Current Body Weight =		170	(pounds)
5	Age, A =		28	(years)
6	Height, H =		68	(inches)
7	Daily Intake, I_0 =		3000	(calories)
8	Activity Level, λ_0 =		1.55	
9				
10	t	W(t)		
11	0	170		
12	1	170.1		
18	7	170.7	1 week	
41	30	172.8	1 month	
194	183	184.9	6 months	
376	365	195.2	1 year	
1836	1825	217.2	5 years	

FIGURE 8.2   Excel output for Exercise 8.2.1.

We use the formula from the text and plug in the relevant parameters:

$$W^* = 0.2205\frac{I_0}{\lambda_0} - 3.5 \cdot H + 1.102 \cdot A - 36.596 \cdot S + 35.494$$
$$= 0.2205\frac{3{,}000}{1.55} - 3.5 \cdot 68 + 1.102 \cdot 28 - 36.596 \cdot 1 + 35.494$$
$$\approx 218.5.$$

Note that this value agrees with our Excel work in Exercise 8.2.1.

5 For parameters of your choosing, set a long-term goal weight.

   a. Determine the daily calorie intake required to reach your long-term goal weight.

   For example purposes we take the parameters from Exercise 8.2.1 and set 160 pounds as the long-term goal weight. We use the formula for W^* from the text to solve for the unknown I_0:

$$W^* = 0.2205\frac{I_0}{\lambda_0} - 3.5 \cdot H + 1.102 \cdot A - 36.596 \cdot S + 35.494$$
$$160 = 0.2205\frac{I_0}{1.55} - 3.5 \cdot 68 + 1.102 \cdot 28 - 36.596 \cdot 1 + 35.494$$
$$160 = 0.1423 I_0 - 208.246$$
$$\frac{368.246}{0.1423} = I_0.$$

We see that the daily calorie intake should equal about 2,588 calories in order to end up with a long-term weight of 160 pounds.

b. Confirm your result in part (a) with Excel.

This is a matter of plugging all parameters into the variable calorie Excel spreadsheet and copying the formulas down until we see the long-term weight appear. In this example, our Excel work produces a long-term weight of 159.9 pounds. The difference can be attributed to rounding error in our calculation of the daily intake.

7   Using your own parameter values and choice of daily intake, compare the long-term predictions of the constant calorie and variable calorie expenditure models. Explain in a complete sentence or two the reasons for any difference.

Using the same parameters as in Exercise 8.2.1, we know that the long-term projection for the variable calorie model will be 218.5 pounds. According to the Mifflin St. Jeor equations, the initial calorie expenditure for the example would be 2,658 calories per day, so we let $E_0 = 2,658$ in the constant calorie model. Since the assumed daily intake is 3,000 calories, the constant calorie model projects a weight that increases to infinity over time.

The difference in the two projections is that the variable calorie expenditure model accounts for increasing calorie needs as people gain weight; eventually calorie needs exactly balance calorie intake and body weight stabilizes.

9   *Extension*: Our variable calorie expenditure model was based on the Mifflin-St. Jeor equations for resting energy expenditure. Construct a variable calorie expenditure model based on the Livingston-Kohlstadt equations for resting energy expenditure instead.

a. Construct an Excel spreadsheet for the model.

First we need to look up the Livingston-Kohlstadt equations, which are based on power law modelling. They are given below:

$$\text{Men}: REE = 208.07 \cdot W^{0.4330} - 5.92 \cdot A$$
$$\text{Women}: REE = 173.27 \cdot W^{0.4536} - 5.09 \cdot A.$$

Here weight, W, is in pounds and age, A, is in years.

We modify the variable calorie expenditure model to use the Livingston-Kohlstadt equations rather than the Mifflin-St. Jeor equations. Once again we make use of an IF statement in Excel to use the correct equation based on the user's sex. Figure 8.3 shows the Excel set-up with the new formula for body weight displayed.

b. Compare the new model's long-term projection for your own weight with the MSJ-based model. Which seems more reasonable?

Figure 8.4 shows a comparison through 5 years of the Livingston-Kohlstadt model and the Mifflin-St. Jeor model for a daily calorie intake of 2,400 calories and a moderate activity level. The long-term projection for the Livingston-Kohlstadt model is about 130.4 pounds and for the Mifflin-St. Jeor model it is about

	A	B	C	D	E	F	G	H	I
1	Variable Calorie Expenditure Model								
2									
3	Sex, S =		M						
4	Current Body Weight =		170	(pounds)					
5	Age, A =		28	(years)					
6	Height, H =		68	(inches)					
7	Daily Intake, I_0 =		2400	(calories)					
8	Activity Level, λ_0 =		1.55						
9									
10	t	$W(t)$			Mifflin-St. Jeor				
11		0	170		170				
12		1	=B11+$C$7/3500-($C$8/3500)*IF($C$3="M",208.07*B11^(0.433)-5.92*$C$5,173.27*						
13		2	B11^(0.4536)-5.09*$C$5)						

FIGURE 8.3 Excel set-up for Exercise 8.2.9.

	A	B	C	D	E
1	Livingston-Kohlstadt Model				
2					
3	Sex, S =		M		
4	Current Body Weight =		170	(pounds)	
5	Age, A =		28	(years)	
6	Height, H =		68	(inches)	
7	Daily Intake, I_0 =		2400	(calories)	
8	Activity Level, λ_0 =		1.55		
9					
10	t		$W(t)$	Mifflin-St. Jeor	
11	0		170	170	
12	1		169.9	169.9	
18	7		169.4 1 week	169.5 1 week	
41	30		167.3 1 month	167.8 1 month	
194	183		156.1 6 months	158.6 6 months	
376	365		146.9 1 year	150.8 1 year	
1836	1825		130.8 5 years	134.1 5 years	

FIGURE 8.4 Excel model comparison for Exercise 8.2.9.

133.2. Both model projections are plausible, though the Livingston-Kohlstadt model is about 3 pounds lighter long-term. Though this will be discussed in the next section, both model projections result in a BMI that is within the "normal" range.

c. Do a general equilibrium analysis for males or females and compare the result to MSJ.

Here we use the equation for females. The DDS for females is given by

$$W(t) = W(t-1) + \frac{I_0}{3,500} - \frac{\lambda_0}{3,500}\left(173.27 \cdot W(t-1)^{0.4536} - 5.09 \cdot A\right).$$

To find the equilibrium value we need to solve for W^*:

$$W^* = W^* + \frac{I_0}{3,500} - \frac{\lambda_0}{3,500}\left(173.27 \cdot (W^*)^{0.4536} - 5.09 \cdot A\right).$$

We get

$$\frac{I_0}{3,500} - \frac{\lambda_0}{3,500}\left(173.27 \cdot \left(W^*\right)^{0.4536} - 5.09 \cdot A\right) = 0$$

$$\frac{\lambda_0}{3,500}\left(173.27 \cdot \left(W^*\right)^{0.4536} - 5.09 \cdot A\right) = \frac{I_0}{3,500}$$

$$173.27 \cdot \left(W^*\right)^{0.4536} - 5.09 \cdot A = \frac{I_0}{\lambda_0}.$$

Continuing to solve for W^* gives

$$173.27 \cdot \left(W^*\right)^{0.4536} = \frac{I_0}{\lambda_0} + 5.09 \cdot A$$

$$\left(W^*\right)^{0.4536} = \frac{I_0}{173.27 \cdot \lambda_0} + 0.0294 \cdot A$$

$$W^* = \left(\frac{I_0}{173.27 \cdot \lambda_0} + 0.0294 \cdot A\right)^{\frac{1}{0.4536}}.$$

We can check our algebra by confirming the result with an example in Excel.

11 Most college students have heard about the "Freshman 15," a term first introduced in an article in *Seventeen* magazine in 1989 for the purported gaining of 15 pounds experienced by students during their first year of college (Karasu, 2013). Many studies have found that the Freshman 15 is a myth (Karasu, 2013). In fact many college students actually lose weight during their first year, and those who do gain weight typically gain somewhere around 3 to 5 pounds, certainly nowhere near the assumed 15 (Karasu, 2013). Moreover, the weight that is gained could simply be due to natural growth and development into adulthood (Posterli, n.d.). Though many factors such as cafeteria-style eating, stress, and irregular sleep habits conspire to make it difficult to eat a healthy diet when starting college, the change most associated with weight gain during this time is an excess of alcohol consumption (Zagorsky & Smith, 2011).

a. Use the variable calorie expenditure model to estimate the required daily calorie intake in order for you to gain 15 pounds in 9 months.

This is a problem we can solve with Goal Seek. We need to find the daily intake, I_0, that results in a weight that is 15 pounds higher than our initial weight around time $t = 275$ days. The Excel set-up just before running Goal Seek is shown in Figure 8.5. After a successful Goal Seek we find that the required daily calorie intake would need to be about $I_0 = 2,907.5$ calories.

b. Use the variable calorie expenditure model to estimate the required daily calorie intake in order for you to gain 3 pounds in 9 months.

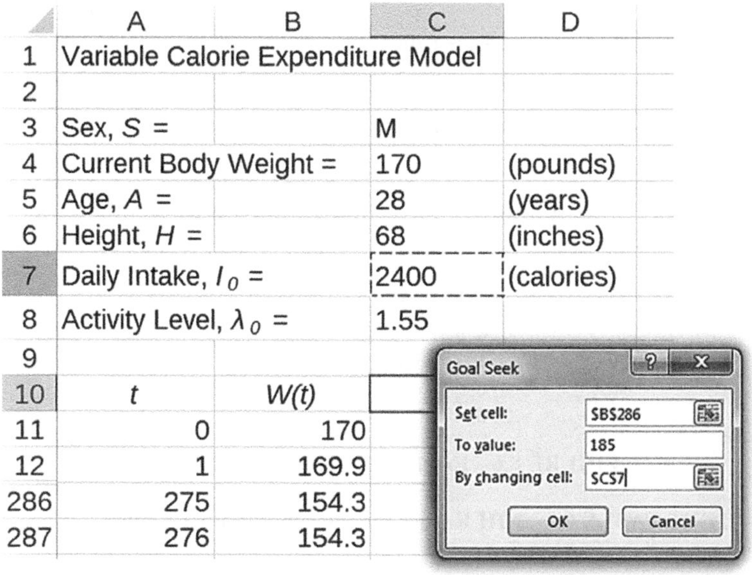

	A	B	C	D
1	Variable Calorie Expenditure Model			
2				
3	Sex, S =		M	
4	Current Body Weight =		170	(pounds)
5	Age, A =		28	(years)
6	Height, H =		68	(inches)
7	Daily Intake, I_0 =		2400	(calories)
8	Activity Level, λ_0 =		1.55	
9				
10	t	$W(t)$		
11	0	170		
12	1	169.9		
286	275	154.3		
287	276	154.3		

Goal Seek

Set cell: $B$286

To value: 185

By changing cell: $C$7

OK   Cancel

FIGURE 8.5   Excel Goal Seek set-up for Exercise 8.2.11.

We re-run Goal Seek with a target weight of 3 pounds over our initial at about time $t = 275$ days. The result is that the required daily calorie intake would need to be about $I_0 = 2,708.9$ calories.

c. Use the variable calorie expenditure model to estimate the amount of weight you would gain in 9 months if you increased your alcohol consumption by 2 drinks per day and changed nothing else.

We assume that we start off with a daily intake that would keep our weight constant. For the given parameter choices this would mean an intake of $I_0 = 2,660$ calories. If we assume that our 2 drinks per day are standard glasses of wine, we need to add about $125 \times 2 = 250$ calories per day to our intake. Then we observe our weight on day $t = 275$. The result is shown in Figure 8.6. The result in this example turns out to be almost exactly equal to the "Freshman 15."

8.3   HEALTH METRICS

1   Find and assess the BMI for a $5'3''$ person who weighs 140 pounds.

We use the formula for BMI from the text to calculate

$$BMI = 703.07 \times \frac{W}{H^2} = 703.07 \times \frac{140}{63^2} \approx 24.8.$$

According to Text Table 8.2, this person's BMI is at the top of the "Normal" range.

◢	A	B	C	D
1	Variable Calorie Expenditure Model			
2				
3	Sex, S =		M	
4	Current Body Weight =		170	(pounds)
5	Age, A =		28	(years)
6	Height, H =		68	(inches)
7	Daily Intake, I_0 =		2910	(calories)
8	Activity Level, λ_0 =		1.55	
9				
10	t	$W(t)$		
11	0	170		
12	1	170.1		
286	275	185.2		
287	276	185.2		

FIGURE 8.6   Excel output for Exercise 8.2.11.

TEXT TABLE 8.2   NIH Ranges for Interpreting BMI Values

BMI	Health Status Category
Below 18.5	Underweight
Between 18.5 and 24.9	Normal
Between 25.0 and 29.9	Overweight
Above 29.9	Obese

3   How tall would a 250-pound person have to be in order to have a healthy BMI?

By "healthy" we understand that we need a BMI in the "Normal" range. That range is from 18.5 to 24.9, so we need to find the heights that would correspond to these two levels. For BMI equal to 18.5 we must solve

$$18.5 = 703.07 \times \frac{250}{H^2}$$
$$18.5 \cdot H^2 = 175,767.5$$
$$H^2 \approx 9500.95.$$

Thus we see that in order for a 250 pound person to have the minimum healthy BMI, the person would need to be about 97.5 inches tall, or a little over 8 feet tall. The calculation for a BMI of 24.9 is similar and yields a required height of about 84 inches, or about 7 feet. This problem illustrates that for an average person a weight of 250 pounds is almost certainly not healthy.

5   Use your own height and the BMI guidelines in Text Table 8.2 to determine a healthy weight range for yourself.

By "healthy" we understand that we need a BMI in the "Normal" range. That range is from 18.5 to 24.9, so we need to find the weights that would correspond to these two levels for a given height. For example purposes we use a height of 5′7″, or 67″. For BMI equal to 18.5 we must solve

$$18.5 = 703.07 \times \frac{W}{67^2}$$
$$18.5 \approx 0.1566 \times W$$
$$118.12 \approx W.$$

Thus we see that in order for a person who is 5′7″ tall to have at least the minimum healthy BMI, the person would need to weigh at least about 118 pounds. The calculation for a BMI of 24.9 is similar and yields a required weight of no more than about 159 pounds. Note that this range of healthy weights is fairly large: about 41 pounds.

7 For parameter values of your choosing, find the projected long-term BMI by using the equilibrium analysis from §8.2.3.

To find the long-term BMI we first find the long-term weight from the equilibrium formula:

$$W^* = 0.2205 \frac{I_0}{\lambda_0} - 3.5 \cdot H + 1.102 \cdot A - 36.596 \cdot S + 35.494.$$

Suppose Theresa is a 24-year-old woman who is 5′4″, consumes 1,750 calories per day and is sedentary. Her long-term weight is given by

$$W^* = 0.2205 \frac{1,750}{1.2} - 3.5 \cdot 64 + 1.102 \cdot 24 - 36.596 \cdot 0 + 35.494$$
$$W^* \approx 159.5.$$

Thus Theresa's long-term BMI will be

$$BMI = 703.07 \times \frac{W^*}{H^2} = 703.07 \times \frac{159.5}{64^2} \approx 27.4.$$

9 *Extension:* Assume that the results of the study referenced at the end of this section hold across all ethnicities. That is, assume that for men, waist size changes by 1″ for every 4.87 pound change in body weight. For women, assume that waist size changes by 1″ for every 5.94 pound change in body weight (Miyatake, Matsumoto, Miyachi, Fujii, & Numata, 2007). For the variable calorie expenditure model, include a new parameter for waist circumference in inches.

a. Have the worksheet compute an initial value for waist to height ratio.

Figure 8.7 shows the Excel set-up with the new parameter and the formula for the initial WHR displayed.

▲	A	B	C	D	E
1	Variable Calorie Expenditure Model with WHR				
2					
3	Sex, S =			M	
4	Current Body Weight =			170	(pounds)
5	Age, A =			28	(years)
6	Height, H =			68	(inches)
7	Waist Size, Wa =			36	(inches)
8	WHR =			=D7/D6	
9	Daily Intake, I_0 =			2400	(calories)
10	Activity Level, λ_0 =			1.55	

FIGURE 8.7   Excel set-up for Exercise 8.3.9a.

b. Have the worksheet model WHR over time by including new columns for waist circumference and WHR next to body weight in the model.

To keep track of waist circumference we need to keep track of how much weight is gained or lost each day. Then we divide the net weight change by 4.87 for men and 5.94 for women to find the corresponding daily change in waist size. The Excel set-up with the formula for waist size displayed is given in Figure 8.8.

▲	A	B	C	D	E
1	Variable Calorie Expenditure Model with WHR				
2					
3	Sex, S =			M	
4	Current Body Weight =			170	(pounds)
5	Age, A =			28	(years)
6	Height, H =			68	(inches)
7	Waist Size, Wa =			36	(inches)
8	WHR =			0.53	
9	Daily Intake, I_0 =			2200	(calories)
10	Activity Level, λ_0 =			1.55	
11					
12	t	$W(t)$	Waist Size(t)	WHR(t)	
13	0	170	36.0	0.53	
14	1	169.9	=C13+(B14-B13)/IF($D$3="M",4.87,5.94)		
15	2	169.7	35.9	0.53	

FIGURE 8.8   Excel set-up for Exercise 8.3.9b.

Finally we calculate WHR over time by referring to the column for waist size. The set-up with the formula for WHR displayed is given in Figure 8.9.

	A	B	C	D	E
1	Variable Calorie Expenditure Model with WHR				
2					
3	Sex, S =			M	
4	Current Body Weight =			170	(pounds)
5	Age, A =			28	(years)
6	Height, H =			68	(inches)
7	Waist Size, Wa =			36	(inches)
8	WHR =			0.53	
9	Daily Intake, I_0 =			2200	(calories)
10	Activity Level, λ_0 =			1.55	
11					
12	t	$W(t)$	$Waist\ Size(t)$	$WHR(t)$	
13	0	170	36.0	0.53	
14	1	169.9	36.0	=C14/$D$6	
15	2	169.7	35.9	0.53	

FIGURE 8.9   Excel set-up for Exercise 8.3.9b continued.

8.4   BODY COMPOSITION

1 Calculate how much body fat and how much lean body mass a person has who weighs 205 pounds and is 29% body fat.

A person who weighs 205 pounds and is 29% body fat will have $205 \times 29\% = 59.45$ pounds of body fat. The rest is lean so the person has $205 - 59.45 = 145.55$ pounds of lean.

3 Estimate your own body fat percentage using the relevant equation by the Navy.

For example purposes we assume Rosa is a female whose measurements are:

$$H = 67''$$
$$Wa = 30''$$
$$Hp = 40''$$
$$N = 13.5''.$$

According to the Navy equation for women, Rosa's body fat percentage is approximately

$$\text{(Women) Body Fat}\% = 163.205 \cdot \log(30 + 40 - 13.5) - 97.684 \cdot \log(67) - 78.387$$
$$\approx 285.941 - 178.378 - 78.387$$
$$\approx 29.2\%.$$

We estimate Rosa's body fat percentage to be 29.2%, which is in the "Acceptable" range as determined by the American Council on Exercise.

5 Estimate your own body fat percentage using the relevant equation by Jackson.

For example purposes we assume Rosa is a 25-year-old female who weighs 145 pounds and is 67″ tall. According to Jackson's equation for women, Rosa's body fat percentage is approximately

$$(\text{Women}) \text{ Body Fat } \% = 0.14{\cdot}A + 39.96{\cdot}\ln\left(703.07\frac{W}{H^2}\right) - 102.01$$

$$= 0.14{\cdot}25 + 39.96{\cdot}\ln\left(703.07\frac{145}{67^2}\right) - 102.01$$

$$\approx 3.5 + 124.79 - 102.01$$

$$\approx 26.3\%.$$

We estimate Rosa's body fat percentage to be 26.3%, which is in the "Acceptable" range as determined by the American Council on Exercise.

7 Estimate your own REE based on the Nelson formula. How does it compare to the MSJ formula for you?

Continuing to use Rosa as an example, at 145 pounds her body fat percentage is 26.3% according to the Jackson equation. Thus Rosa has $145 \times 26.3\% = 38.135$ pounds of body fat and $145 - 38.135 = 106.865$ pounds of lean. The Nelson formula estimates her REE to be

$$REE = 1.832{\cdot}F + 11.708{\cdot}L$$

$$= 1.832{\cdot}38.135 + 11.708{\cdot}106.865$$

$$\approx 1,321 \text{ calories per day.}$$

Recall that Rosa is 67″ tall and 25 years old. The Mifflin-St. Jeor equation estimates her REE to be

$$REE = 4.536{\cdot}W + 15.875{\cdot}H - 5{\cdot}A + 166{\cdot}S - 161$$

$$= 4.536{\cdot}145 + 15.875{\cdot}67 - 5{\cdot}25 + 166{\cdot}0 - 161$$

$$\approx 1,435 \text{ calories per day.}$$

The MSJ equation estimates Rosa's REE to be about 100 calories more than Nelson's equation.

9 Suppose Dianne weighs 155 pounds and is 24% body fat. She incurs a calorie deficit of 3500 calories over the course of a week. Estimate how much fat, lean, and total body weight Dianne should expect to lose during the week.

First we need to determine how much body fat Dianne has. We get $155 \times 24\% = 37.2$ pounds of body fat for Dianne. Using the energy partition equations from the text we find the proportions of her calorie deficit devoted to fat and lean to be

$$\text{Proportion devoted to fat} = \frac{F}{F+6} = \frac{37.2}{43.2} \approx 0.86$$

$$\text{Proportion devoted to lean} = \frac{6}{F+6} = \frac{6}{43.2} \approx 0.14.$$

Assuming the change in these proportions is negligible during the week, we calculate the numbers of calories burned from fat and from lean to be

$$\text{calories of fat burned} = 0.86 \times 3,500 = 3,010$$
$$\text{calories of lean burned} = 0.14 \times 3,500 = 490.$$

To calculate the weight change for fat and lean we use their energy densities:

$$\text{pounds of fat burned} = \frac{3,010}{4,279.43} \approx 0.703$$

$$\text{pounds of lean burned} = \frac{490}{823.38} \approx 0.595.$$

We see that Dianne will lose approximately 0.7 pounds of fat and 0.6 pounds of lean body mass during the week. Her total weight loss expected is thus about 1.3 pounds.

11   *Extension*: Create an Excel spreadsheet for calculating body fat percentage based on the Navy's equations. Make judicious use of "IF" statements to make the spreadsheet as user friendly as possible.

Figure 8.10 shows the Excel set-up with the formula for body fat percentage displayed. As a check on our Excel work we enter the parameters for Rosa from Exercise 8.4.3 and note that Figure 8.11 shows we get the same result.

	A	B	C	D	E	F	G	H
1	Estimating Body Fat Using Navy Equations							
2								
3	Sex, S =			F				
4	Height, H =			67	(inches)			
5	Waist Size, Wa =			30	(inches)			
6	Neck circumference, N =			13.5	(inches)			
7	Hips (women only), Hp =			40	(inches)			
8								
9	Body Fat % Estimate:		=IF(D3="M",86.01*LOG(D5-D6)-70.041*LOG(D4)+36.76,163.205*LOG(D5+D7-D6)-					
10			97.684*LOG(D4)-78.387)					

FIGURE 8.10   Excel set-up for Exercise 8.4.11.

	A	B	C	D	E
1	Estimating Body Fat Using Navy Equations				
2					
3	Sex, S =			F	
4	Height, H =			67	(inches)
5	Waist Size, Wa =			30	(inches)
6	Neck circumference, N =			13.5	(inches)
7	Hips (women only), Hp =			40	(inches)
8					
9	Body Fat % Estimate:		29.2		

FIGURE 8.11   Excel output for Exercise 8.4.11.

8.5   THE BODY COMPOSITION MODEL FOR BODY WEIGHT

1   Eduardo is 20 years old, 5′ 9″ tall, and weighs 165 pounds. He is highly active and consumes 2200 calories per day. Project Eduardo's body weight and body fat percentage 1 week, 1 month, 6 months, 1 year, and 5 years from today.

We use the body composition Excel spreadsheet we developed in this section. We input Eduardo's parameters and copy the formulas down to at least $t = 1825$ days, or 5 years. By referring to the appropriate cells we can display the body weight and body fat projections for all required times in Figure 8.12.

	A	B	C	D	E	F	G	H	I
1	Body Composition Model								
2									
3	Sex, S =	M			Initial BMI =		24.4		
4	Current Body Weight =	165		(pounds)	Initial Body Fat % =		18.0%	(Jackson)	
5	Age, A =	20		(years)					
6	Height, H =	69		(inches)			Results After:	Weight	Body Fat %
7	Daily Intake, I_0 =	2200		(calories)			1 week	163.3	17.7%
8	Activity Level, λ_0 =	1.725					1 month	157.9	16.7%
9							6 months	133.2	12.3%
10	t	$F(t)$	$L(t)$	$W(t)$	Body Fat %		1 year	122.3	10.6%
11	0	29.7	135.3	165.0	18.0%		5 years	119.0	10.1%
12	1	29.6	135.2	164.8	18.0%				
13	2	29.5	135.0	164.5	17.9%				
14	3	29.3	134.9	164.3	17.9%				

FIGURE 8.12   Excel output for Exercise 8.5.1.

3   Use a daily calorie intake and activity level of your choice and the Excel body composition model to project your own body composition and body weight for 1 week, 1 month, 6 months, 1 year, and 5 years from today.

For example purposes we assume Rosa is a 25-year-old female who weighs 145 pounds and is 67″ tall. Her measurements are

$$Wa = 30''$$
$$Hp = 40''$$
$$N = 13.5''$$
$$T = 24''$$
$$C = 13.5''$$
$$Wr = 5.5''.$$

a. Run the projections using Jackson's estimate for your initial body fat percentage.

We assume Rosa is extremely active and consumes 2,250 calories per day. Using the body composition spreadsheet yields the projections shown in Figure 8.13.

	A	B	C	D	E	F	G	H	I
1	Body Composition Model								
2									
3	Sex, S =		F		Initial BMI =		22.7		
4	Current Body Weight =		145	(pounds)	Initial Body Fat % =		26.3%	(Jackson)	
5	Age, A =		25	(years)					
6	Height, H =		67	(inches)			Results After:	Weight	Body Fat %
7	Daily Intake, I_0 =		2250	(calories)			1 week	144.3	26.1%
8	Activity Level, λ_0 =		1.9				1 month	142.3	25.7%
9							6 months	132.8	23.9%
10	t	$F(t)$	$L(t)$	$W(t)$	Body Fat %		1 year	127.6	22.8%
11	0	38.1	106.9	145.0	26.3%		5 years	124.4	22.2%
12	1	38.0	106.9	144.9	26.3%				
13	2	38.0	106.8	144.8	26.2%				
14	3	37.9	106.8	144.7	26.2%				

FIGURE 8.13 Excel output for Exercise 8.5.3a.

b. Run the projections using the appropriate Navy equation to estimate your initial body fat percentage.

According to the Navy equation for women, Rosa's body fat percentage is approximately

$$(\text{Women}) \text{ Body Fat \%} = 163.205 \cdot \log(30 + 40 - 13.5) - 97.684 \cdot \log(67) - 78.387$$
$$\approx 285.941 - 178.378 - 78.387$$
$$\approx 29.2\%.$$

Using 29.2% as Rosa's initial body fat percentage changes the projections of the model. The new projections are given in Figure 8.14.

	A	B	C	D	E	F	G	H	I
1	Body Composition Model								
2									
3	Sex, S =		F		Initial BMI =		22.7		
4	Current Body Weight =		145	(pounds)	Initial Body Fat % =		29.2%		
5	Age, A =		25	(years)					
6	Height, H =		67	(inches)			Results After:	Weight	Body Fat %
7	Daily Intake, I_0 =		2250	(calories)			1 week	144.6	29.1%
8	Activity Level, λ_0 =		1.9				1 month	143.2	28.8%
9							6 months	136.6	27.6%
10	t	$F(t)$	$L(t)$	$W(t)$	Body Fat %		1 year	132.8	26.8%
11	0	42.3	102.7	145.0	29.2%		5 years	129.8	26.2%
12	1	42.3	102.6	144.9	29.2%				
13	2	42.3	102.6	144.9	29.2%				
14	3	42.2	102.6	144.8	29.2%				
15	4	42.2	102.6	144.7	29.1%				

FIGURE 8.14 Excel output for Exercise 8.5.3b.

c. Run the projections using the appropriate Bailey equation to estimate your initial body fat percentage.

According to the Covert Bailey equation for women under 30, Rosa's body fat percentage is approximately

$$(\text{Women 30 or younger}) \text{ Body Fat } \% = Hp + 0.8{\cdot}T - 2{\cdot}C - Wr$$
$$= 40 + 0.8{\cdot}24 - 2{\cdot}13.5 - 5.5$$
$$= 26.7\%.$$

Using 26.7% as Rosa's initial body fat percentage changes the projections of the model. The new projections are given in Figure 8.15.

	A	B	C	D	E	F	G	H	I
1	Body Composition Model								
2									
3	Sex, S =		F		Initial BMI =		22.7		
4	Current Body Weight =		145	(pounds)	Initial Body Fat % =		26.7%		
5	Age, A =		25	(years)					
6	Height, H =		67	(inches)			Results After:	Weight	Body Fat %
7	Daily Intake, I_0 =		2250	(calories)			1 week	144.4	26.6%
8	Activity Level, λ_0 =		1.9				1 month	142.4	26.2%
9							6 months	133.4	24.4%
10	t	$F(t)$	$L(t)$	$W(t)$	Body Fat %		1 year	128.4	23.4%
11	0	38.7	106.3	145.0	26.7%		5 years	125.2	22.7%
12	1	38.7	106.2	144.9	26.7%				
13	2	38.6	106.2	144.8	26.7%				
14	3	38.6	106.2	144.7	26.6%				
15	4	38.5	106.1	144.6	26.6%				

FIGURE 8.15   Excel output for Exercise 8.5.3c.

d. Discuss how much of a difference in long-term weight projection results from using the different body fat estimates.

The 5-year weight projections range from 124.4 pounds to 129.8 pounds, and the body fat projections range from 22.2% to 26.2%.

5 Show that the long-term values for body fat and lean mass you found for yourself in Exercise 8.5.3 satisfy the equilibrium equation $I_0 - \lambda_0(1.832{\cdot}F^* + 11.708{\cdot}L^*) = 0$.

We continue to use Rosa as our example. In Exercise 8.5.3 we have three different long-term projections based on the three different estimates for initial body fat percentage. We plug the projections for fat and lean for each of these into the equilibrium equation

$$2,250 - 1.9{\cdot}(1.832{\cdot}F^* + 11.708{\cdot}L^*) = 0.$$

For an initial body fat of 26.3%, the long-term projections for fat and lean are 27.6 and 96.8 pounds, respectively. Thus we have

$$2,250 - 1.9{\cdot}(1.832{\cdot}27.6 + 11.708{\cdot}96.8) \approx 2,250 - 1.9{\cdot}1,183.9$$
$$\approx 0.$$

The other confirmations are similar, and the minor differences from zero are due to rounding error.

7   We often use the estimate that 1 pound of body weight contains approximately 3500 calories. Use the Excel body composition model to estimate how many calories would be contained in a one-pound change of your own body weight.

For example purposes, we assume John is a 6′ tall, lightly active 22-year-old male who weighs 200 pounds and is 27% body fat. Thus John carries 54 pounds of body fat and 146 pounds of lean. We know from Nelson's equation that John initially will burn

$$1.375 \cdot [1.832 \cdot 54 + 11.708 \cdot 146] \approx 2,486.4$$

calories per day.

Next we use Goal Seek to find how many calories per day would be required for John to lose 1 pound of body weight in 1 week. The Excel set-up just before running Goal Seek is given in Figure 8.16. The result of the Goal Seek is a daily intake requirement of about 2,053. This intake is $2,486.4 - 2,053 = 433.4$ calories per day below John's initial requirement. Over the course of the week John therefore incurs a calorie deficit of about $7 \times 433.4 = 3,033.8$ total calories and loses exactly one pound of body weight. Thus our estimate is that initially for John a one-pound change in body weight contains about 3,034 calories.

	A	B	C	D	E	F
1	Body Composition Model					
2						
3	Sex, S =		M		Initial BMI =	
4	Current Body Weight =		200	(pounds)	Initial Body Fat % =	
5	Age, A =		22	(years)		
6	Height, H =		72	(inches)		
7	Daily Intake, I_0 =		2486.4	(calories)		
8	Activity Level, λ_0 =		1.375			
9						
10	t	$F(t)$	$L(t)$	$W(t)$		
11	0	54.0	146.0	200.0		
12	1	54.0	146.0	200.0	27.0%	
13	2	54.0	146.0	200.0	27.0%	
14	3	54.0	146.0	200.0	27.0%	
15	4	54.0	146.0	200.0	27.0%	
16	5	54.0	146.0	200.0	27.0%	
17	6	54.0	146.0	200.0	27.0%	
18	7	54.0	146.0	200.0	27.0%	

Goal Seek

Set cell: D18
To value: 199
By changing cell: SCS7

OK   Cancel

FIGURE 8.16   Excel Goal Seek set-up for Exercise 8.5.7.

9   *Extension:* Set a 2-year weight goal and use the body composition Excel model to find the daily calorie intake necessary for you to achieve your goal. Using this daily calorie intake, answer the following.

a. What percentage of the long-term weight change takes place in the first month?

For example purposes we assume that Claire is a 5′3″ female who is 20 years old, highly active, and weighs 135 pounds. She is currently 20% body fat, and her goal weight is 125 pounds. Using Goal Seek we find that in order for her weight to be 125 pounds on day 730 Claire needs to consume about 2,138 calories per day. With her intake set to 2,138 we see that Claire is projected to weigh 133.5 pounds in 1 month, or on day 30. Thus out of the 10 total pounds she wants to lose, 1.5 of them will be lost during the first month, or about 15% of the weight loss will occur during the first month.

b. What percentage of the long-term weight change takes place in the first 6 months?

Looking at day 182 we see that Claire is projected to weigh 128.6 pounds. Thus during the first 6 months she loses 6.4 pounds of the desired 10. She loses 64% of the desired weight during the first 6 months.

c. What percentage of the long-term weight change takes place in the first year?

Looking at day 365 we see that Claire is projected to weigh 126.2 pounds. Thus during the first year she loses 8.8 pounds of the desired 10. Thus she loses 88% of the desired weight during the first year.

8.6 POINTS-BASED SYSTEMS: THE WEIGHT WATCHERS™ MODEL

1 Use a nutrition facts label of your choice to find the total calories and total adjusted calories for a serving of the food.

According to www.sabra.com, one 2-tablespoon serving of Sabra's classic hummus contains 5 g of fat, 2 g of protein, 4 total grams of carbohydrate, and 1 g of dietary fiber. Thus the total number of calories in a serving is given by

$$TC = 4 \cdot P + 4 \cdot (C - DF) + 4 \cdot DF + 9 \cdot F$$
$$= 4 \cdot 2 + 4 \cdot (4 - 1) + 4 \cdot 1 + 9 \cdot 5$$
$$= 69.$$

Our result agrees within rounding error with the 70 calories per serving reported on the label.

To find the total adjusted calories, we must account for the fact that the body absorbs different proportions of each macronutrient. Using the equation developed in the text we get

$$TAC = 3.2 \cdot P + 3.8 \cdot (C - DF) + DF + 9 \cdot F$$
$$= 3.2 \cdot 2 + 3.8 \cdot (4 - 1) + 1 + 9 \cdot 5$$
$$= 63.8.$$

Thus out of the approximately 70 calories per serving of the hummus, only about 64 calories will actually be absorbed by the body after digestion.

3 Find the PointsPlus® value for a serving of the food in Exercise 8.6.1.

The PointsPlus value for a serving of the Sabra hummus is $\frac{TAC}{35} = \frac{63.8}{35} \approx 1.82 \approx 2$.

5 Find the daily PointsPlus target for Doug, a 200-pound man who is $5'10''$ tall and 22 years old.

We use the Weight Watchers PointsPlus allotment spreadsheet and plug in Doug's parameters. The set-up and result are shown in Figure 8.17. We see that Doug will have 46 PointsPlus to spend on a daily basis. If we assume that Doug will use his 49 weekly points, then his total daily points allotment is in effect 53.

	A	B	C	D
1	Weight Watchers™ PointsPlus® Allotments			
2				
3	Sex, S =		M	
4	Current Body Weight =		200	(pounds)
5	Age, A =		22	(years)
6	Height, H =		70	(inches)
7				
8	Daily Calorie Needs, DC =			3094.6
9	Adjusted Daily Calorie Needs, ADC =			2785.1
10	Daily PointsPlus® Allotment =			46

FIGURE 8.17 Excel output for Exercise 8.6.5.

7 Kara weighs 120 pounds, and she exercises at high intensity for 30 min. How many Activity Points does she earn?

Using the equation from the text for high intensity exercise we get

$$\text{Points at high intensity} = 0.000808 \times \text{minutes} \times W$$
$$= 0.000808 \times 30 \times 120$$
$$\approx 2.91 \approx 3.$$

Kara will earn 3 Activity Points for her exercise.

9 How many minutes of moderate intensity exercise would it take for you to earn 4 Activity Points?

We assume a body weight of 270 pounds, and we take advantage of rounding so that we need to solve:

$$\text{Points at moderate intensity} = 0.000327 \times \text{minutes} \times W$$
$$3.5 = 0.000327 \times \text{minutes} \times 270.$$

We get

$$\frac{3.5}{0.000327 \cdot 270} = \text{minutes}$$
$$39.64 \approx \text{minutes}.$$

We would need to exercise for a minimum of about 40 min at moderate intensity in order to earn 4 Activity Points.

11   Set your own weight goal and use the Excel Weight Watchers model to determine how long it will take you to reach it if a) you do not exercise, and b) you exercise for a length of time and intensity of your choosing.

We use the dynamic Weight Watchers spreadsheet developed in the section and input the required parameters. Once the correct parameters are entered, we copy the formulas down until the body weight reaches our goal. We do this once with 0 entered for the daily minutes of exercise, and once with the exercise parameter of our choosing entered. We should reach the goal weight faster with exercise than without.

13   Once you meet your goal weight in Exercise 8.6.11, determine how many points you should consume per day to maintain your weight. Assume no activity points.

What is required here is to have the Weight Watchers spreadsheet calculate your daily calorie needs at your goal weight. Once you have the daily calorie needs, divide that number by 38.9 (the assumed number of calories per PointsPlus) to get your maintenance point level.

15   Suppose you exercise each day for a length of time and intensity of your choosing. What should your new maintenance PointsPlus level be?

Each Activity Point is equivalent to about 77.8 calories burned. This means that each Activity Point is really equivalent to 2 PointsPlus if we are only interested in maintaining weight. Take the number of Activity Points generated by your chosen exercise parameters and double it. Then add that to the maintenance level you found in Exercise 8.6.13 to get the total number of daily points you can use and still maintain your goal weight.

Printed and bound by CPI Group (UK) Ltd, Croydon, CR0 4YY

27/10/2024

14580475-0005